BROOKLANDS BOOKS

LAND ROVER DISCOVERY
1989 - 1994

Compiled by
R.M.Clarke

ISBN 1 85520 231X

Brooklands Books Ltd.
PO Box 146, Cobham, KT11 1LG
Surrey, England

Printed in Hong Kong

BROOKLANDS BOOKS

BROOKLANDS ROAD TEST SERIES

Abarth Gold Portfolio 1950-1971
AC Ace & Aceca 1953-1983
Alfa Romeo Giulietta Gold Portfolio 1954-1965
Alfa Romeo Giulia Berlinas 1962-1976
Alfa Romeo Giulia Coupés 1963-1976
Alfa Romeo Giulia Coupés Gold P. 1963-1976
Alfa Romeo Spider 1966-1990
Alfa Romeo Spider Gold Portfolio 1966-1991
Alfa Romeo Alfasud 1972-1984
Alfa Romeo Alfetta Gold Portfolio 1972-1987
Alfa Romeo Alfetta GTV6 1980-1987
Allard Gold Portfolio 1937-1959
Alvis Gold Portfolio 1919-1967
American Motors Muscle Cars 1966-1970
Armstrong Siddeley Gold Portfolio 1945-1960
Aston Martin Gold Portfolio 1972-1985
Austin Seven 1922-1982
Austin A30 & A35 1951-1962
Austin Healey 100 & 100/6 Gold P. 1952-1959
Austin Healey 3000 Gold Portfolio 1959-1967
Austin Healey Sprite 1958-1971
BMW Six Cyl. Coupés 1969-1975
BMW 1600 Collection No.1 1966-1981
BMW 2002 Gold Portfolio1968-1976
BMW 316, 318, 320 (4 cyl.) Gold P. 1975-1990
BMW 320, 323, 325 (6 cyl.) Gold P .1977-1990
BMW 5 Series Gold Portfolio1981-1987
BMW M Series Performance Portfolio1976-1993
Bristol Cars Gold Portfolio 1946-1992
Buick Automobiles 1947-1960
Buick Muscle Cars 1965-1970
Cadillac Automobiles 1949-1959
Cadillac Automobiles 1960-1969
Chevrolet 1955-1957
Chevrolet Impala & SS 1958-1971
Chevrolet Corvair 1959-1969
Chevy El Camino & SS 1959-1987
Chevy II Nova & SS 1962-1973
Chevelle & SS Muscle Portfolio 1964-1972
Chevrolet Muscle Cars 1966-1971
Chevy Blazer 1969-1981
Chevrolet Corvette Gold Portfolio 1953-1962
Chevrolet Corvette Sting Ray Gold P. 1963-1967
Chevrolet Corvette Gold Portfolio 1968-1977
High Performance Corvettes 1983-1989
Camaro Muscle Portfolio 1967-1973
Chevrolet Camaro Z28 & SS 1966-1973
Chevrolet Camaro & Z28 1973-1981
High Performance Camaros 1982-1988
Chrysler 300 Gold Portfolio 1955-1970
Chrysler Valiant 1960-1962
Citroen Traction Avant Gold Portfolio 1934-1957
Citroen 2CV Gold Portfolio 1948-1989
Citroen DS & ID 1955-1975
Citroen DS & ID Gold Portfolio 1955-1975
Citroen SM 1970-1975
Cobras & Replicas 1962-1983
Shelby Cobra Gold Portfolio 1962-1969
Cobras & Cobra Replicas Gold P. 1962-1989
Cunningham Automobiles 1951-1955
Daimler SP250 Sports & V-8 250 Saloon Gold Portfolio 1959-1969
Datsun Roadsters 1962-1971
Datsun 240Z 1970-1973
Datsun 280Z & ZX 1975-1983
The De Lorean 1977-1993
De Tomaso Collection No. 1 1962-1981
Dodge Charger 1966-1974
Dodge Muscle Cars 1967-1970
Dodge Viper on the Road
The De Lorean 1977-1993
Excalibur Collection No. 1 1952-1981
Facel Vega 1954-1964
Ferrari Cars 1946-1956
Ferrari Collection No. 1 1960-1970
Ferrari Dino 1965-1974
Ferrari Dino 308 1974-1979
Ferrari 308 & Mondial 1980-1984
Motor & T&CC Ferrari 1966-1976
Motor & T&CC Ferrari 1976-1984
Fiat Pininfarina 124 & 2000 Spider 1968-1985
Fiat-Bertone X1/9 1973-1988
Ford Consul, Zephyr, Zodiac Mk.I & II 1950-1962
Ford Zephyr, Zodiac, Executive, Mk.III & Mk.IV 1962-1971
Ford Cortina 1600E & GT 1967-1970
High Performance Capris Gold P. 1969-1987
Capri Muscle Portfolio 1974-1987
High Performance Fiestas 1979-1991
High Performance Escorts Mk.I 1968-1974
High Performance Escorts Mk.II 1975-1980
High Performance Escorts 1980-1985
High Performance Escorts 1985-1990
High Performance Sierras & Merkurs Gold Portfolio 1983-1990
Ford Automobiles 1949-1959
Ford Fairlane 1955-1970
Ford Ranchero 1957-1959
Thunderbird 1955-1957
Thunderbird 1958-1963
Thunderbird 1964-1976
Ford Falcon 1960-1970
Ford GT40 Gold Portfolio 1964-1987
Ford Bronco 1966-1977
Ford Bronco 1978-1988
Holden 1948-1962
Honda CRX 1983-1987
Hudson & Railton 1936-1940
Isetta 1953-1964
Jaguar and SS Gold Portfolio 1931-1951
Jaguar XK120, 140, 150 Gold P. 1948-1960
Jaguar Mk.VII, VIII, IX, X, 420 Gold P.1950-1970
Jaguar 1957-1961
Jaguar Mk.2 1959-1969
Jaguar Cars 1961-1964
Jaguar E-Type Gold Portfolio 1961-1971
Jaguar E-Type 1966-1971
Jaguar E-Type V-12 1971-1975
Jaguar XJ12, XJ5.3, V12 Gold P. 1972-1990
Jaguar XJ6 Series II 1973-1979
Jaguar XJ6 Series III 1979-1986
Jaguar XJS Gold Portfolio 1975-1990
Jeep CJ5 & CJ6 1960-1976
Jeep CJ5 & CJ7 1976-1986
Jensen Cars 1946-1967
Jensen Cars 1967-1979
Jensen Interceptor Gold Portfolio 1966-1986
Jensen Healey 1972-1976
Lagonda Gold Portfolio 1919-1964
Lamborghini Cars 1964-1970
Lamborghini Countach & Urraco 1974-1980
Lamborghini Countach & Jalpa 1980-1985
Lancia Beta Gold Portfolio 1972-1984
Lancia Fulvia Gold Portfolio 1963-1976
Lancia Stratos 1972-1985
Land Rover Series I 1948-1958
Land Rover Series II & IIa 1958-1971
Land Rover Series III 1971-1985
Land Rover 90 & 110 1983-1989
Land Rover Discovery 1989-1994
Lincoln Gold Portfolio 1949-1960
Lincoln Continental 1961-1969
Lincoln Continental 1969-1976
Lotus & Caterham Seven Gold P. 1957-1993
Lotus Sports Racers Gold Portfolio 1953-1965
Lotus Elite 1957-1964
Lotus Elite & Eclat 1974-1982
Lotus Elan Gold Portfolio 1962-1974
Lotus Elan Collection No. 2 1963-1972
Lotus Elan 1989-1992
Lotus Cortina Gold Portfolio 1963-1970
Lotus Europa Gold Portfolio 1966-1975
Lotus Turbo Esprit 1980-1986
Motor & T&CC on Lotus 1979-1983
Marcos Cars 1960-1988
Maserati 1965-1970
Maserati 1970-1975
Mazda RX-7 Collection No. 1 1978-1981
Mercedes Benz Cars 1949-1954
Mercedes Benz Competition Cars 1950-1957
Mercedes Benz Cars 1954-1957
Mercedes Benz Cars 1957-1961
Mercedes 190 & 300 SL 1954-1963
Mercedes 230/250/280SL 1963-1971
Mercedes Benz SLs & SLCs Gold P. 1971-1989
Mercedes S & 600 1965-1972
Mercedes S Class 1972-1979
Mercury Muscle Cars 1966-1971
Metropolitan 1954-1962
MG Gold Portfolio 1929-1939
MG TC 1945-1949
MG TD 1949-1953
MG TF 1953-1955
MGA & Twin Cam Gold Portfolio 1955-1962
MG Midget Gold Portfolio1961-1979
MGB Roadsters 1962-1980
MGB MGC & V8 Gold Portfolio 1962-1980
MGB GT 1965-1980
Mini Cooper Gold Portfolio 1961-1971
Mini Muscle Cars 1961-1979
Mini Moke Gold Portfolio1964-1994
Mopar Muscle Cars 1964-1967
Morgan Three-Wheeler Gold Portfolio 1910-1952
Morgan Plus 4 & Four 4 Gold P. 1936-1967
Morgan Cars 1960-1970
Morgan Cars Gold Portfolio 1968-1989
Morris Minor Collection No. 1 1948-1980
Shelby Mustang Muscle Portfolio 1965-1970
High Performance Mustang IIs 1974-1978
High Performance Mustangs 1982-1988
Oldsmobile Automobiles 1955-1963
Oldsmobile Cutlass & 4-4-2 1964-1972
Oldsmobile Muscle Cars 1964-1971
Oldsmobile Toronado 1966-1978
Opel GT 1968-1973
Packard Gold Portfolio 1946-1958
Pantera Gold Portfolio 1970-1989
Panther Gold Portfolio 1972-1990
Plymouth Barracuda 1964-1974
Plymouth Muscle Cars 1966-1971
Pontiac Tempest & GTO 1961-1965
Pontiac Muscle Cars 1966-1972
Pontiac Firebird & Trans-Am 1973-1981
High Performance Firebirds 1982-1988
Pontiac Fiero 1984-1988
Porsche 356 1952-1965
Porsche 911 1965-1969
Porsche 911 1970-1972
Porsche 911 1973-1977
Porsche 911 Carrera 1973-1977
Porsche 911 Turbo 1975-1984
Porsche 911 SC 1978-1983
Porsche 914 Collection No. 1 1969-1983
Porsche 914 Gold Portfolio 1969-1976
Porsche 924 Gold Portfolio 1975-1988
Porsche 928 1977-1989
Porsche 944 Gold P.1981-1991
Range Rover Gold Portfolio 1970-1992
Reliant Scimitar 1964-1986
Riley Gold Portfolio 1924-1939
Riley 1.5 & 2.5 Litre Gold Portfolio 1945-1955
Rolls Royce Silver Cloud & Bentley 'S' Series Gold Portfolio 1955-1965
Rolls Royce Silver Shadow Gold P. 1965-1980
Rover P4 1949-1959
Rover P4 1955-1964
Rover 3 & 3.5 Litre Gold Portfolio 1958-1973
Rover 2000 & 2200 1963-1977
Rover 3500 1968-1977
Rover 3500 & Vitesse 1976-1986
Saab Sonett Collection No.1 1966-1974
Saab Turbo 1976-1983
Studebaker Gold Portfolio 1947-1966
Studebaker Hawks & Larks 1956-1963
Avanti 1962-1990
Sunbeam Tiger & Alpine Gold P. 1959-1967
Toyota MR2 1984-1988
Toyota Land Cruiser 1956-1984
Triumph TR2 & TR3 1952-1960
Triumph TR4, TR5, TR250 1961-1968
Triumph TR6 Gold Portfolio 1969-1976
Triumph TR7 & TR8 Gold Portfolio 1975-1982
Triumph Herald 1959-1971
Triumph Vitesse 1962-1971
Triumph Spitfire Gold Portfolio 1962-1980
Triumph 2000, 2.5, 2500 1963-1977
Triumph GT6 1966-1974
Triumph Stag 1970-1980
TVR Gold Portfolio 1959-1990
VW Beetle Gold Portfolio1935-1967
VW Beetle Gold Portfolio1968-1991
VW Beetle Collection No.1 1970-1982
VW Karmann Ghia 1955-1982
VW Bus, Camper, Van 1954-1967
VW Bus, Camper, Van 1968-1979
VW Bus, Camper, Van 1979-1989
VW Scirocco 1974-1981
VW Golf GTI 1976-1986
Volvo PV444 & PV544 1945-1965
Volvo Amazon-120 Gold Portfolio 1956-1970
Volvo 1800 Gold Portfolio 1960-1973

BROOKLANDS ROAD & TRACK SERIES

Road & Track on Alfa Romeo 1949-1963
Road & Track on Alfa Romeo 1964-1970
Road & Track on Alfa Romeo 1971-1976
Road & Track on Alfa Romeo 1977-1989
Road & Track on Aston Martin 1962-1990
R & T on Auburn Cord and Duesenburg 1952-84
Road & Track on Audi & Auto Union 1952-1980
Road & Track on Audi & Auto Union 1980-1986
Road & Track on Austin Healey 1953-1970
Road & Track on BMW Cars 1966-1974
Road & Track on BMW Cars 1975-1978
Road & Track on BMW Cars 1979-1983
R & T on Cobra, Shelby & Ford GT40 1962-1992
Road & Track on Corvette 1953-1967
Road & Track on Corvette 1968-1982
Road & Track on Corvette 1982-1986
Road & Track on Corvette 1986-1990
Road & Track on Datsun Z 1970-1983
Road & Track on Ferrari 1975-1981
Road & Track on Ferrari 1981-1984
Road & Track on Ferrari 1984-1988
Road & Track on Fiat Sports Cars 1968-1987
Road & Track on Jaguar 1950-1960
Road & Track on Jaguar 1961-1968
Road & Track on Jaguar 1968-1974
Road & Track on Jaguar 1974-1982
Road & Track on Jaguar 1983-1989
Road & Track on Lamborghini 1964-1985
Road & Track on Lotus 1972-1981
Road & Track on Maserati 1952-1974
Road & Track on Maserati 1975-1983
R & T on Mazda RX7 & MX5 Miata 1986-1991
Road & Track on Mercedes 1952-1962
Road & Track on Mercedes 1963-1970
Road & Track on Mercedes 1971-1979
Road & Track on Mercedes 1980-1987
Road & Track on MG Sports Cars 1949-1961
Road & Track on MG Sports Cars 1962-1980
Road & Track on Mustang 1964-1977
R & T on Nissan 300-ZX & Turbo 1984-1989
Road & Track on Peugeot 1955-1986
Road & Track on Pontiac 1960-1983
Road & Track on Porsche 1951-1967
Road & Track on Porsche 1968-1971
Road & Track on Porsche 1972-1975
Road & Track on Porsche 1975-1978
Road & Track on Porsche 1979-1982
Road & Track on Porsche 1982-1985
Road & Track on Porsche 1985-1988
R & T on Rolls Royce & Bentley 1950-1965
R & T on Rolls Royce & Bentley 1966-1984
Road & Track on Saab 1972-1992
R & T on Toyota Sports & GT Cars 1966-1984
R & T on Triumph Sports Cars 1953-1967
R & T on Triumph Sports Cars 1967-1974
R & T on Triumph Sports Cars 1974-1982
Road & Track on Volkswagen 1951-1968
Road & Track on Volkswagen 1968-1978
Road & Track on Volkswagen 1978-1985
Road & Track on Volvo 1957-1974
Road & Track on Volvo 1975-1985
R&T - Henry Manney at Large & Abroad

BROOKLANDS CAR AND DRIVER SERIES

Car and Driver on BMW 1955-1977
Car and Driver on BMW 1977-1985
C and D on Cobra, Shelby & Ford GT40 1963-84
Car and Driver on Corvette 1956-1967
Car and Driver on Corvette 1968-1977
Car and Driver on Corvette 1978-1982
Car and Driver on Corvette 1983-1988
C and D on Datsun Z 1600 & 2000 1966-1984
Car and Driver on Ferrari 1955-1962
Car and Driver on Ferrari 1963-1975
Car and Driver on Ferrari 1976-1983
Car and Driver on Mopar 1956-1967
Car and Driver on Mopar 1968-1975
Car and Driver on Mustang 1964-1972
Car and Driver on Pontiac 1961-1975
Car and Driver on Porsche 1955-1962
Car and Driver on Porsche 1963-1970
Car and Driver on Porsche 1970-1976
Car and Driver on Porsche 1977-1981
Car and Driver on Porsche 1982-1986
Car and Driver on Saab 1956-1985
Car and Driver on Volvo 1955-1986

BROOKLANDS PRACTICAL CLASSICS SERIES

PC on Austin A40 Restoration
PC on Land Rover Restoration
PC on Metalworking in Restoration
PC on Midget/Sprite Restoration
PC on Mini Cooper Restoration
PC on MGB Restoration
PC on Morris Minor Restoration
PC on Sunbeam Rapier Restoration
PC on Triumph Herald/Vitesse
PC on Spitfire Restoration
PC on Beetle Restoration
PC on 1930s Car Restoration

BROOKLANDS HOT ROD 'MUSCLECAR & HI-PO ENGINES' SERIES

Chevy 265 & 283
Chevy 302 & 327
Chevy 348 & 409
Chevy 350 & 400
Chevy 396 & 427
Chevy 454 thru 512
Chrysler Hemi
Chrysler 273, 318, 340 & 360
Chrysler 361, 383, 400, 413, 426, 440
Ford 289, 302, Boss 302 & 351W
Ford 351C & Boss 351
Ford Big Block

BROOKLANDS RESTORATION SERIES

Auto Restoration Tips & Techniques
Basic Bodywork Tips & Techniques
Basic Painting Tips & Techniques
Camaro Restoration Tips & Techniques
Chevrolet High Performance Tips & Techniques
Chevy Engine Swapping Tips & Techniques
Chevy-GMC Pickup Repair
Chrysler Engine Swapping Tips & Techniques
Custom Painting Tips & Techniques
Engine Swapping Tips & Techniques
Ford Pickup Repair
How to Build a Street Rod
Land Rover Restoration Tips & Techniques
MG 'T' Series Restoration Guide
Mustang Restoration Tips & Techniques
Performance Tuning - Chevrolets of the '60s
Performance Tuning - Pontiacs of the '60s

BROOKLANDS MILITARY VEHICLES SERIES

Allied Military Vehicles No.1 1942-1945
Allied Military Vehicles No.2 1941-1946
Complete WW2 Military Jeep Manual
Dodge Military Vehicles No.1 1940-1945
Hail To The Jeep
Land Rovers in Military Service
Off Road Jeeps: Civ. & Mil. 1944-1971
US Military Vehicles 1941-1945
US Army Military Vehicles WW2-TM9-2800
VW Kubelwagen Military Portfolio1940-1990
WW2 Jeep Military Portfolio 1941-1945

261033

CONTENTS

5	Grooming a Star	*Land Rover Owner*	Nov.	1989
8	On the Trail of the Discovery	*Autocar & Motor*	Sept. 13	1989
15	Promoting the Product	*Land Rover Owner*	Jan.	1990
18	Discovery Shows Japan the Way	*Autocar & Motor*	Nov. 15	1989
20	Discovery v Shogun v Trooper Comparison Test	*Autocar & Motor*	Nov. 15	1989
28	Inside Story	*Autocar & Motor*	Nov. 15	1989
30	Voyage of Discovery	*What Car?*	Jan.	1990
38	Land Rover Discovery V8 Road Test	*Autocar & Motor*	Jan. 3	1990
41	New Doors Open for Discovery	*Autocar & Motor*	Sept. 12	1990
42	A Drovers Tale	*Performance Car*	July	1990
48	Land Rover Discovery V8i 5dr	*Autocar & Motor*	Oct. 17	1990
52	Discovery v Land Cruiser v Shogun v Range Rover - Will the Japanese Ever Win Comparison Test	*Car*	Nov.	1990
60	Fashion Wagons - Discovery v Land Cruiser v Shogun Comparison Test	*What Car?*	Jan.	1991
64	Hard at Work Running Report	*Autocar & Motor*	July 3	1991
65	Bog Standards - Discovery - Landcruiser - G-Wagen Comparison Test	*Autocar & Motor*	Feb. 6	1991
71	Land Rover Discovery V8i - Tried	*Fast Lane*	Feb.	1991
72	Land Rover Discovery Tdi Long Term Test	*Autocar & Motor*	Apr. 3	1991
76	Land Rover Discovery	*Modern Motor*	May	1991
78	Discovery Under Fire	*Autocar & Motor*	July 31	1991
80	Gallic Trial Discovery Test	*Land Rover Owner*	Aug.	1991
82	Discovery V8i 5dr	*Car South Africa*	Sept.	1991
88	Running Amuck	*Car*	Apr.	1992
92	Discovery V8i Automatic	*Land Rover Owner*	July	1993
94	White Collar Dirt	*Autocar & Motor*	July 14	1993
96	Discovery Auto Tdi	*Land Rover Owner*	Oct.	1993
98	Less is More for New Discovery	*What Car?*	July	1993
99	Urban Cowboy Road Test	*Land Rover Owner*	Oct.	1993

ACKNOWLEDGEMENTS

Land Rover's Discovery has been a success from the moment it was announced in 1989, and it has generated such interest that owners have already started clamouring for a Brooklands Book about it. We are obviously only too pleased to oblige with this 100-pager in our regular Road Test series, which covers the models available during the first four years of production.

We are pleased to record our gratitude for their co-operation to those who own the copyright to the original material reproduced in this book. Our thanks go to the managements of *Autocar and Motor, Car, Car South Africa, Fast Lane, Land Rover Owner, Modern Motor, Performance Car* and *What Car?*. Our thanks also go to motoring writer James Taylor for the few words of introduction which follow, and we recommend to readers his own book on the Discovery, *Land Rover Discovery - the Enthusiast's Companion*, published by Motor Racing Publications.

R M Clarke

At the beginning of the 1980's Land Rover Ltd was faced with falling demand for its products. Many of the company's traditional markets in developing countries had been lost to cheaper Japanese utility vehicles, and the arrival of a new breed of leisure-market 4x4's from Japan had also created a lucrative new market sector in the developed countries. Recognising that it would have to follow the market in order to survive, Land Rover decided to develop a new leisure-market model to beat the Japanese makers at their own game.

The Discovery was that model. Based on the acclaimed chassis of the Range Rover and using much of that vehicle's powertrain, the Discovery took the market by storm after its introduction at the Frankfurt Motor Show in 1989. It immediately became a best seller, outselling the best of the Japanese imports in its home country and in 1993 going on sale with Honda badges in Japan - perhaps the ultimate admission by the Land of the Rising Sun that its motor manufacturers could not build a better vehicle.

Right from the beginning, Land Rover knew that it could not rest on its laurels because the Japanese makers were improving their products all the time. So within a year of the original model's launch came a five-door alternative to the three-door body and a more powerful petrol engine. Further improvements pushed the Discovery gradually up-market so that its most expensive variants were nudging the bottom of the Range Rover's price bracket, and at the same time the range was expanded downwards, most obviously with the introduction of a smaller-engined model in 1993.

Clearly, this is only the beginning. As this book goes to press, the Discovery has been on the market for barely four years, and there can be absolutely no doubt that Land Rover Ltd. has further exciting developments in the pipeline. Among those expected is the launch of the vehicle into the vast US market, where it will follow in the footsteps of its big brother, the Range Rover. This collection of road tests and other reports on the Discovery makes clear what a powerful impact the vehicle has had on the four-wheel-drive leisure market, and comes thoroughly recommended to owners and would-be owners alike.

James Taylor

▲ 'Body drop' operations are completed as the body meets the chassis. Radiator is fitted and brake connections made.

Grooming a star

OF COURSE you could write a book about it (and someone will) - the story behind the launch of Discovery. Land Rover had a lot on its mind in August and early September but Tom Sheppard managed to get a feel for what goes into the launch of a new model.

YOU GET used to new launches in the motoring press. First the grainy spy pictures, then the comment from 'informed sources'. Then comments from clearly uninformed sources, then a few ads, then the release shots and a road test or two. It is happening all the time.

But when it is, so to speak, in the family - it is different. As the leaks were published I couldn't help (having some experience of major engineering projects and the heartache they cause) willing it all to go well, feeling for the targets being set, the inevitable hiccups in development, the agonising decisions have to be made over timescales, costs, outside suppliers, performance, the need for a particular fix and the hideous implications it could have in tooling costs.

The production people need long lead times to see the new line organised and built and for suppliers to get their act together. The engineers and design people need the maximum time they can get to make sure things are right. The conflicting needs have somehow to be reconciled. At some stage someone has to freeze the design and say 'That is what we build!' And at the same time they must be realistic enough to allow some flexibility in case last minute changes really do have to be implemented.

Development time

That is why the average development time for any new model is around five years, longer in many cases. And it is why Land Rover's achievement of completing the process on Discovery in under three is all the more remarkable.

If we think at all about what is involved in a launch, we probably think of a press release, some advertisements, some brochures and half a dozen vehicles for the press to play around with. It isn't really like that at all. In the case of Discovery the vehicle count for the press fleet is no less than eighty. Getting off to a good start is nowhere more important than in the car industry and on these vehicles will ride the reputation of the new model for a considerable time to come - remember rust and the Vauxhall Victor!. And, in the case of an on-off road 4×4 where, dare one say it, the standard of specialist knowledge among most automotive writers is perhaps less than comprehensive, the impression made by 'the car bits' is doubly crucial.

After Frankfurt the premier for Discovery (1990 Range Rover was presented at the end of September as well, for good measure), the factory must support a five week-long press facility for driving impressions to be gained. Such facilities are a story in themselves with considerable midnight oil burned by the devoted band of multi-role engineer/mechanic/demonstrator/valet jacks-of-all-trades who contrive, each day of the event, to have 20 or more apparently 'new' vehicles available - no matter how muddy the conditions had been the day before. Something like 200 journalists will be involved with Discovery in mid October. Later, for a fewer number, off-road impressions can be sampled by the cognoscenti at Eastnor. Then there will be the loan vehicles going out . . .

And that is just the UK. Launch will be in Italy around the same time and in other European countries after that. And each wretched journalist has to be made to feel he is very special. If the articles we have seen about the Range Rover in recent years is anything to go by, it is odds-on that splashing through muddy puddles for the photographer is the average magazine's idea of off-roading and its comprehensive of centre diff-locks, viscous couplings, articulation and the like - the nuts and bolts of real off-roading performance - will be hazy.

By Tom Sheppard

Groo
a

▲ *Body wiring is installed before body reaches main track.*

▼ *Dash goes in pre-assembled during trim assembly phase. Internal trim is highly tooled and contributes to outstanding smoothness of internal appearance.*

▼ *Front seats are installed using power tools set to required torques. Twin pipes are for heater ducting to rear compartment.*

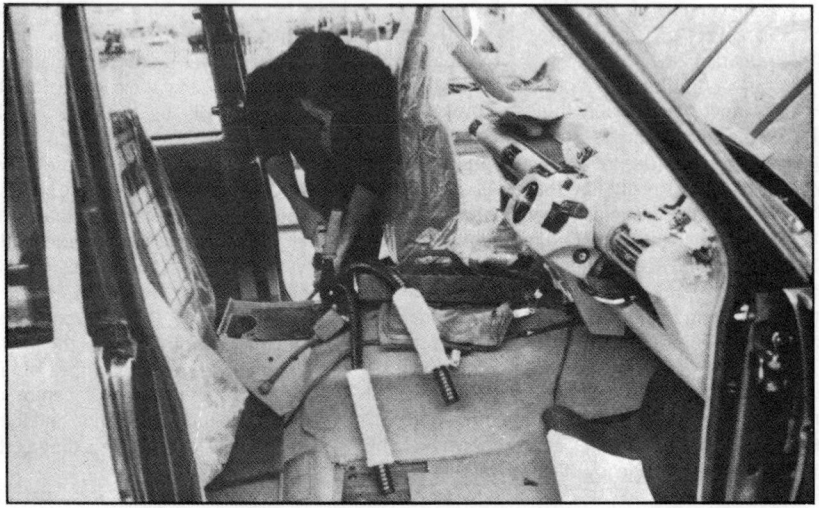

The public relations people, the marketing department, the film and stills people, the writers and brochure designers will have been briefed, their wares seen and refined and the material will have been ready in advance of Frankfurt and the first press faclity - if only just! The show stand had to be designed, the press conference had to be organised, the exhibitions people had to cope with a million details - hopefully in advance, but as like as not on the spot and in a foreign language! (To keep the organisation on its toes, the hotel where the Frankfurt press conference was held and for which invitations had been issued, changed its name three days before the show!)

Making it

And, of course, someone has actually to make the vehicles. When I was at Solihull the Discovery line was in steady operation though not up to full speed. There were some four hundred vehicles in the despatch compound; at sales launch in mid November there will be 2500 Discoveries available for delivery. The engineers will have thought out long ago about how the machine will be assembled, the back of a thousand envelopes scribbled on. But one bright morning the assembly workers are producing - carefully and very slowly at first - Discoveries instead of Land Rovers or Range Rovers.

Land Rover have a new production concept, TQM - Total Quality Management - which aims to use the perceptions and engineering experience of *everyone* in the production process rather than just that of the inspectors. No one accepts work from a colleague unless it is up to par. No 'fix it later' postponements are accepted. It makes for a slow start but it makes for sustained and soundly based quality for the end products. Production by year's end will be about 300 per week.

In the press workshop there were about forty Discoveries - some hard-worked development vehicles, some obviously part of what was to be the launch fleet and being meticulously valetted by what must have been the most priviledged YTS lads in the country — getting an inside look at what many would have paid money to get a view of. Many vehicles - these were obviously early models - were being modded up to production standard and had entrails of instrument binnacles hanging out.

Photography is not a quick business and Simon Maris of Land Rover Public Relations had obviously come top of the 'being-patient-with-photographers' course. Mervyn Rogers and Mal (who drove the vehicle on the LRO October cover) seemed to have similar qualifications. Tracking shots are as notoriously difficult to get

Discovery Launch

ming
:ar

reliably as are panning pictures. Slow shutter speeds have to be used to get the blur of the road and background but slow shutter speeds over a test track as rough as the one at Solihull is a guarantee of severe camera shake. Patience is required on all sides; and a lot of film - and a time. And there will have been rafts of other writers and photographers to cope with.

Early research

Land Rover claim that Discovery has been exceptionally well researched and talking to the 'Project J' project director, Mike Donovan, the claim seems well justified. Born as a conept just before Christmas 1986, research was intensive and the focus on the particular market sector equally so.

Splitting the on/off road firmament into three separate market sectors - luxury, utility, and, in the middle, what they called the 'personal transport sector' - Land Rover started from scratch with customer expectations and aspirations, detailed studies of usage, reactions to vehicles currently on the market place in this sector and the needs of those whom they term 'conquests from the car sector'. People who want a bit more than just the space of an estate; people who want - Land Rover cringe at the cliché but admit it is nevertheless a valid term - a 'lifestyle vehicle'.

Their studies also reveal that the total 4×4 market has grown 250 per cent in the past five years. Of this market, personal transport category sales that were 25,000 in 1983 topped 100,000 in 1988. And Land Rover estimate this will double again in the next five years. Their investment in development tooling and the new production line totals £100 million and - another oft-forgotten aspect of 'grooming a star' - dealers have matched that figure in provision of new showrooms and workshops facilities. The new vehicle, Land Rover point out, is 'incremental to' not 'instead of' the Range Rover and Land Rover line.

Direction of the Discovery project has clearly been broad and extremely incisive with simultaneous inputs from all specialist sectors right from the earliest days. Test criteria in terms of climatic extremes and durability have been no less severe for Discovery than those applicable to the hard working 'utility' Land Rovers. Technical triumphs and sagas, unsung in an age when we are increasingly dependent on technology, are something our ephemeral media fail daily to pick up on. Getting a small glimpse of what went on, in so short a timescale, to produce Discovery leaves me with (as it may other readers of LRO) no small feeling of admiration.

▲ 'Body-drop is the meeting of body with chassis
▼ Body picks up One Ten mounting points as in Range Rover: Note excellent load spreading design of spare wheel carrier.

ON THE TRAIL

Discovery is Land Rover's first new model since the Range Rover in 1970. Its mission is to make or break Land Rover over the next decade. Howard Lees looks under the skin of the new car

OF DISCOVERY

conquer the leisure market dominated by the Mitsubishi Shogun and its success or failure will
and meets the men who are staking their reputations on it. Photography by Derek Goard

Discovery gets the interior it needs to compete with Japanese, including 60/40 split rear bench (above) and optional extra rear seats which fold away when not in use (right). Discovery promises to be outstanding off-roader

Land Rover's first new model for almost two decades was unveiled yesterday at the Frankfurt show. Discovery — the second choice of name after the preferred Highlander tag was seen on a Volvo off-road truck — is a new Range Rover-sized off-roader that fills the huge price and specification gap between the Land Rover and Range Rover. It will compete head-on with the Mitsubishi Shogun and Isuzu Trooper and be crucial to Land Rover's survival well into the '90s.

Available initially in three-door form only — a five door version will follow in mid-1990 — Discovery retains the traditional Range Rover benefits of permanent four-wheel drive, coil-sprung suspensions and panels of aluminium alloy. Under the bonnet it comes with either a carburettored version of the long-serving all-alloy 3.5-litre V8, or Land Rover's own brand new 2.5-litre direct injection turbo diesel, the 200 Tdi.

As revealed exclusively by *Autocar & Motor* last year (24 February 1988). Discovery sits on the Range Rover's 100ins wheelbase, but is 2ins longer overall and boasts a great deal more interior space. A near vertical side-hinged tailgate with the spare wheel slung on the outside, combined with a higher roof line behind the front seats, contributes to a luggage capacity of 46 cubic feet, 10 more than the Range Rover can muster. This is enough to house a pair of fold-down seats, giving the vital option of space for seven adults.

Sitting on a Range Rover-type box-section steel ladder chassis is a steel bodyshell to which aluminium alloy panels are attached — although the stepped roof is in steel. Despite the use of alloy, Discovery is no lightweight — the Tdi's kerbweight of 4551lb is over 600lb more than a Shogun 5-door TD. A large glass area includes alpine lights in the sides of the raised rear roof, and the option of two glass sunroofs — the front removable to be stowed in a special pocket on the back of the rear seat.

Running gear is from the Range Rover parts bin. Front and rear live axles have the same 3.54 final drive ratio and 58.5ins track as the Range Rover. Coil springs, progressive rate at the rear and devoid of anti-roll bars, are Land Rover's usual recipe for long wheel travel and fine articulation. The familiar leading and trailing radius arms locate the axles fore and aft, with a Panhard rod at the front and an upper 'A' frame at the rear preventing lateral movement. Steering is power-assisted recirculating ball.

Discovery sits on 7ins wide pressed-steel Range Rover wheels, shod with Goodyear Wranglers rather than the usual Michelin XM+S 205R16s. The all-disc braking system, with a drum parking brake acting on the transfer box output, is also pure Range Rover.

Providing the permanent four-wheel drive that none of the competition can match is Land Rover's LT230T dual-range transfer box, complete with manually-lockable centre differential. It's the unit that was fitted to the Range Rover until the new Borg Warner viscous coupling-equipped unit was introduced for all 1989 models, and gives Discovery a clear traction edge on wet or slippery roads — conditions in which part-time 4wd vehicles with no centre diff have to soldier through with only two driven wheels.▶

John Bilton

GAMBLE PAYS OFF

"THE REAL KEY, THE THING THAT we have been so impressed with, is the fuel consumption." John Bilton, currently Rover's manager for powertrain strategic planning but the man in charge of powertrain engineering at Land Rover when the 200 Tdi was developed, is clearly delighted with Discovery's new turbo diesel. And with the diesel version expected to outsell the V8 petrol 3:1, to compete with the Japanese the engine had to deliver the goods.

Work began in 1985, before plans for Discovery were finalised. "We knew we needed to update and uprate our diesel engine, and the choice was whether to go the direct or indirect route." At that time direct injection technology on small, high speed engines was by no means as highly-developed as it is today, although it had demonstrated significant fuel economy benefits.

In particular, combustion noise and refinement were known to be problem areas and the decision to go direct injection was something of a gamble. Bilton was convinced that work on two-spring injectors being carried out by Bosch and Lucas/CAV would provide the key: "We knew the industry was working on two-spring injectors, and were confident that by late '86 a solution would have been found."

Bosch injectors were eventually chosen for the 200 Tdi, although during development Land Rover worked very closely with CAV as well. Effectively the two-spring system provides a rising-rate return spring for the injectors, slowing the initial rate of injection and controlling the rate of rise of combustion pressure and hence the characteristic diesel 'knock' — especially on start-up.

But a successful high speed di engine also depends on a very highly developed combustion chamber. For the technology it needed, Land Rover went to AVL in Graz, Austria. "AVL are really the world's leading diesel engine consultants and have eight to 10 years' experience of direct injection. It was their research that provided the combustion process used in the 200 Tdi, and they also advised on durability and reliability."

Development was carried out in-house at Land Rover, its experience of turbocharging gained from the existing turbo diesel proving invaluable on the intercooled Tdi. "The turbocharger matching is remarkable. It has been a steep learning curve, but the engineers here are now ahead of the game by quite a margin." The 1800rpm peak torque figure and early reports of superb driveability with minimal turbo lag back this up.

The engine shares dimensions and is based around Land Rover's existing 2.5 litre turbo diesel, although all its component parts are new. Block, pistons, conrods and of course the aluminium alloy cylinder head are unique to the Tdi — the cylinder block actually has the direction of water flow through it reversed, but some clever design work allows it to be machined on the same line as the existing casting. Cylinder heads are produced on a separate, brand-new CNC machining centre. "About the only item that is common is the crankshaft — although we have improved the cold-rolling process and put the new crank back in the old engine."

Lighter by 45lb than the old turbo diesel, with more torque at much lower rpm and almost as much power as the VM unit used in the Range Rover, clearly the 200 Tdi would be suitable for both Land Rover and Range Rover as well — in fact over 2 million road test miles have already been racked up on the new engine in both. "Obviously the engine has potential in other products — these things don't stand still. But expanding production of the Tdi would mean another CNC line for cylinder heads, and that is why any future application has yet to be decided."

Bilton is defensive about the choice of the venerable 3.5-litre V8 as the petrol unit, pointing to its fantastic reputation for durability and performance, if a little surprised by its thirst. "I was expecting an urban cycle fuel figure of about 15mpg, and I don't know why it only returned 13mpg. Of course there is potential for a smaller petrol engine, but you have to justify the cost."

Even the development of Land Rover's own diesel engine for Discovery was by no means certain. "There was quite a lot of pressure during the development of the 200 Tdi to opt for a bought-in engine. The final decision was only made about nine months ago. Even if the Tdi had not gone into production, the knowledge we gained of turbocharging, direct injection and two-spring injectors would have been worth it."

The verdict will have to wait until we drive it, but all the signs are that John Bilton was right to fight for Land Rover's own engine.

◀ There is no automatic gearbox for Discovery, the demand in this sector of the market being minuscule, so both V8 and turbo diesel drive through the Range Rover LT77 five-speeder. Internal ratios are unchanged, except for a lower first gear on the turbo diesel only. If needed, the ZF four-speed automatic offered in the Range Rover could be engineered to fit.

None of this breaks new mechanical ground — indeed it is the same recipe that has kept the Range Rover on top since 1970. But under the bonnet of the turbo diesel — expected to be by far the most popular model — is a brand new 2.5-litre direct injection engine, so far unique outside the world of commercial vehicles.

The Land Rover-developed 200 Tdi shares bore centres and its 90.47mm × 97mm bore and stroke with Land Rover's existing 2.5-litre turbo diesel, but the casting itself is new. Topping the iron block is a new aluminium alloy cylinder head, its two valves per cylinder still pushrod operated but featuring fuel injection directly into the combustion chambers. Bosch two-spring injectors are used to keep combustion noise — always the Achilles heel of direction units — to a minimum and avoid the characteristic clatter on cold-start.

Turbocharged and intercooled like the 2.4-litre VM unit used in the Range Rover Turbo D, the Tdi develops 111bhp at 4000rpm. That is 1bhp down on the Italian unit but produced 200rpm lower. Far more impressive is the bottom end torque — a hefty 195lb ft at just 1800rpm compared with the peaky VM's 183lb ft at 2400rpm.

With that huge spread of torque, fine specific power and the efficiency of direct injection. Discovery stacks up very well indeed. A top speed of 92mph and 0-60mph time of 16.2secs heads the competition by some margin, and is backed up by best-in-class fuel consumption figures — 30.5mpg at 75mph.

Petrol models get the venerable 3.5-litre Rover V8, these days owned and built exclusively by Land Rover. In V8 Land Rover's twin-SU carb guise, the engine gets a higher compression ratio to run unleaded rather than two-star fuel and a different camshaft. The bottom line is 145bhp at 5000rpm and 192lb ft of torque at 2800rpm, enough to propel Discovery to a claimed top speed of 102mph and to 60mph in 12.7secs. No catalyst version will be available until the fuel-injected version is introduced next year.

You pay heavily for the V8's extra performance at the pumps — the V8 returns just 13mpg at 75mph. At least a fuel capacity of 19.5 galls should ensure a decent range between fills.

The dated, cramped and downright uncomfortable interior has prevented the top-of-the-range Land Rover County from competing against the Japanese. All that has changed with Discovery, and while it doesn't have the leather and walnut opulence of a Range Rover, it can hold its head up with the best.

Supportive-looking front seats, a 60/40 split rear bench — fold-down rear seats are optional — and carpet throughout create a thoroughly modern and well-laid out interior. There are stowage bins everywhere. There's the usual crop of instruments and controls — speedo, rev counter, clock, fuel and temperature gauges, and a welcome freedom from gimmicks like inclinometers and altimeters. A Clarion radio/cassette is standard, but the optional audio pack includes a more upmarket unit with remote control.

Other options include the extra rear seats, air

Driving environment is well thought out and free from gimmicks (above). Engine (right) is brand-new 2.5-litre direct injection turbo diesel — unique in a non-commercial vehicle. Bosch two-spring injectors keep combustion noise to a minimum

conditioning and an electric pack which includes electric windows and mirrors, central locking and headlamp washers. A factory fitted roof rack is available which can be detached and stowed under the rear seat.

For once, Land Rover has been determined to get the jump on aftermarket suppliers, and a 50-strong range of accessories including nudge bars, winches, ski-carriers and extra lights will be on offer immediately.

On paper, Discovery has the credentials to blow the opposition out of the water, backed up by aggressive pricing that should see the Solihull product actually undercut the best-selling Mitsubishi Shogun.

It will be hampered initially by the lack of a five-door model or an injected petrol engine that can be fitted with a catalyst, but both those problems will be addressed next year. Discovery is only Land Rover's third-ever product, but it looks like continuing the legend of the Land Rover and Range Rover.

LAND ROVER DISCOVERY TD

LAYOUT
Longitudinal front engine/four-wheel drive

ENGINE
Capacity 2495cc, 4 cylinders in line.
Bore 90.47mm, **stroke** 97mm.
Compression ratio 19.5 to 1.
Head/block al. alloy/cast iron.
Valve gear 2 valves per cylinder.
Fuel and ignition mechanical two-spring diesel injection.
Max Power 111bhp (PS Din) (83 kW ISO) at 4000rpm.
Max Torque 195lb ft (205Nm) at 1800rpm.

GEARBOX
Five-speed manual, dual range transfer box. **Ratios:** Top 0.77; 4th 1.00; 3rd 1.397; 2nd 2.132; 1st 3.692. Final drive ratio 3.538, 25.1mph/1000rpm in top. Transfer ratio: high 1.222, low 3.320.

SUSPENSION
Front live axle, radius arms, Panhard rod, coil springs, telescopic dampers.
Rear live axle, radius arms, upper 'A' frame, coil springs, telescopic dampers.

STEERING
Recirculating ball, 3.4 turns lock to lock, power assisted.

BRAKES
Front 11.8ins (298mm) discs **Rear** 11.4ins (290mm) discs.

WHEELS AND TYRES
Pressed steel, 7ins rims, Goodyear Wrangler 205R16 tyres.

DIMENSIONS
Length   178ins (4521mm)
Width   70.6ins (1793mm)
Height   75.6ins (1919mm)
Wheelbase   100ins (2540mm)
Weight   4551lb (2080kg)

PERFORMANCE (claimed)
0-60mph (100km/h)   16.8secs
Maximum speed   92mph

FUEL CONSUMPTION
30.5mpg urban, 42.3mpg constant 56mph, 28.9mpg constant 75mph. Fuel tank 19.5gals (88.6 litres).

LAND ROVER DISCOVERY V8

As Turbo Diesel except for

ENGINE
Capacity 3528cc, 8 cylinders in 90deg vee.
Bore 89.9mm, **stroke** 71.1mm.
Compression ratio 8.1 to 1.
Head/block al. alloy/al. alloy.
Fuel and ignition electronic ignition, twin SU carburettors.
Max Power 145bhp (PS Din) (108 kW ISO) at 5000rpm.
Max Torque 192lb ft (260Nm) at 2800rpm.

GEARBOX
Ratios: 1st 3.321.

DIMENSIONS
Weight   4359lb (1932kg)

FUEL CONSUMPTION
13.0mpg urban, 26.2mpg constant 56mph, 19.6mpg constant 75mph.

Chris Woodwark

TAKING ON JAPAN

THE MAN RESPONSIBLE FOR TURNing Discovery into the success that Land Rover needs is commercial director Chris Woodwark. It won't be the first time he's overseen a launch that could make or break a company — he was sales and marketing director of Leyland Trucks when the Roadrunner 7.5-tonner was introduced in 1983, and commercial director of Austin Rover at the launch of the 800 in 1986.

"Discovery is the most important launch for Land Rover since 1970, when we introduced the Range Rover," he says. "It sits squarely between Land Rover and Range Rover, and is aimed directly at certain competitors. For the first time it gives us an opportunity to compete with Japan."

Discovery was given the green light just two and a half years ago, and it will be aimed at a completely different market to Land Rover or Range Rover. "It is a leisure vehicle, not aimed at the luxury sector at all. Discovery, if you like, is for Yuppies and Range Rover is for people who've already made it."

Woodwark is clear about Discovery's strengths. "The first thing that will sell Discovery is the Land Rover badge. Secondly, it rides and handles just like a Range Rover — I don't expect Discovery owners to do much off-roading, but the capability is there." And Woodwark is convinced that the new interior will win over a lot of buyers: "It's so user-friendly."

He doesn't deny that the company was slow to react to the enormous growth of the 4wd leisure market, a sector so dominated by the Japanese that it wouldn't be unfair to call it the Shogun class. "The first priority was to sort out the cost side of the Land Rover business. We rationalised 12 sites in to one which cost a lot of money at the time but now saves us a great deal. We used to cover over a million miles a year just transporting components between plants. At the same time, we had to develop our products to make money."

Woodwark cites the moving of the Range Rover upmarket as the fruits of that policy — after which the company was able to enter North America very successfully. At the same time, the company's parts operation was rationalised, but only then could Land Rover devote any resources to a third product line.

The metamorphosis of Range Rover was so successful that production now easily outstrips the combined Land Rover Ninety and One Ten output. At present, the Discovery production line is turning out 200 cars a week and, if it's successful, could overtake Range Rover. "We are aiming to take a significant share of a market worth one third of a million cars a year."

Woodwark claims to be happy with the profitability of Land Rover now, and production of the vehicle that spawned the company will continue at Solihull for the foreseable future. "We tried to drag the Land Rover into the leisure market with the County, but it wasn't very successful. Now we will concentrate on commercial, military and agricultural markets. But Land Rover *will* be developed further — just watch this space."

A trip around Land Rover's Solihull plant reveals a full range of competitive 4wds under evaluation, including smaller vehicles like the Suzuki Vitara, but Woodwark isn't planning a fourth product line. "I don't see Land Rover ever approaching the smaller 4wd market directly, because you need a lot of car technology. For a £10,000 vehicle like this to make sense, you need production runs of 50-100,000 vehicles."

In the meantime, Discovery will be launched with a massive publicity and marketing offensive. According to marketing director John Russell, the vast majority of customers will not have experienced a Land Rover product before. "About 25 per cent of customers in this sector have had a small 4wd, and trade up to a larger off-roader. But virtually all the business will be incremental. Discovery won't really affect sales of Land Rover County or Range Rover."

Central to this campaign will be a TV commercial, featuring Discovery in the snow and mountains of New Zealand. And an intensive programme to generate awareness will be run through Land Rover's 130 dealers.

Discovery will be a three door only at its launch, a decision Russell defends passionately: "A three-door Discovery was the best strategy to get us into the sector at the right time, and allowed us to come in price-competitive with our rivals." No one at Land Rover is yet saying what price-competitive means, but in the UK a three-door Shogun turbo diesel costs £15,239 and a five-door £17,399. Expect a base Tdi Discovery to be somewhere in between.

DISCOVERY LAUNCH

▲ Across the water Plymouth slumbers while the Discovery fleet assembles at Mount Edgecumbe ready for the next day's guests to arrive by boat.

◀ A hall at Plymouth's bus garage was taken over as a flowline for cleaning and repairing the Discovery launch fleet.

Promoting

OF EQUAL importance to getting the product right is a successful launch. Mike McCabe sees how this was achieved at Land Rover's greatest show ever — the Discovery launch.

THOSE WITH a good product are at pains to ensure the quality of the product is matched by the quality of its presentations. Thus it is at Land Rover's events for the public, press or dealers.

Not since 1970 had the company had a completely new product line, and one on which its fortunes will depend into the twenty-first century. So it's no surprise that all the stops were pulled out for the European launch of Discovery at Plymouth.

It was a return to the West Country for Land Rover, because in 1970 their second product line, the Range Rover, was launched near Falmouth in Cornwall. By contrast with Plymouth, that was a small affair; a mere 20 demonstration vehicles and a modest country hotel.

Various sites were considered at home and abroad for Discovery's send-off, but in the end it was Plymouth that could offer all the necessary facilities. Moreover, the British product had a British venue pleasing to home visitors and especially so to guests from abroad who were in the majority. The maritime setting also served well in the presentations because of thoughts of Scott of the Antarctic's vessel "Discovery", and the Pilgrim Fathers setting sail from Plymouth on their voyage of discovery.

Following the launches of the Range Rover and Series 3 in the early '70s, there was product stagnation and the Land Rover launch almost became an extinct species. But with the formation of Land Rover Ltd in 1978 and an injection of cash, fortunes began to rise. The following year saw the first fruits with the arrival of the V8 109" Land Rover and an event at Eastnor Castle estate entitled "Land Rover through the '80s."

Continuing development have made

Making a splash. Dealers and journalists were able to sample some gentle off-roading across Dartmoor

Some of the unsung heroes who worked day and night for five weeks to clean and repair the Discovery launch fleet.

the product

Report and pictures by Mike McCabe

Land Rover launches an established part of the motoring scene, and a decade of progress has been brought to a fitting climax with the launch of the Discovery.

Jetting in

For five weeks in October and November, journalists and dealers arrived at Plymouth in separate groups and stayed one or two nights at the Moat House Hotel on Plymouth Hoe. This was the venue for business meetings and presentations, and was the base for Land Rover's event administration.

While guests were jetting-in to Exeter airport on some forty special flights, work went on at three sites around Plymouth to make sure the launch ran well.

At Millbay Docks a surprise was in store for the dealers. This site housed a specially built presentation suite and theatre in a most unlikely looking cargo warehouse. The inside had been transformed into a high-tech show of sound, light and movement. This was the first sight the dealers had of the new product which arrived on a turntable in a cloud of carbon dioxide.

A multi-projector display showed the vehicle in action and gave information to back the speakers. Principle speaker was John Russell, sales and marketing director, and when foreign guests were involved, simultaneous translation was given into languages via radio headphones.

Suitably fired with enthusiasm, the dealers found the door through which they had entered had disappeared, and another had opened which led them directly onto the waterfront. From here boats took them across Plymouth Sound to collect the fleet of Discoveries which were lined up on the water's edge at Mount Edgcumbe estate.

They then set out in the vehicles to drive a scenic route in Cornwall and Devon, stopping at St. Mellion Golf Club for lunch, and at Blanchford Manor on the edge of Dartmoor for a traditional Devonshire cream tea. On the way they took in some

DISCOVERY LAUNCH

DISCOVERY LAUNCH

Roger Crathorne, Land Rover's presentations team manager sets the Bruff guidance on the truck.

(Far right): It's important to fit the steering wheel lock before taking to the rails.

▲ *John Russell, Land Rover's marketing director addresses international dealers in the specially constructed theatre at Millbay Docks.*

mild cross-country driving on private land.

On journalist days, the driving was a similarly scenic route, but the starting point was Plymouth Hoe parade, and lunch was taken at the Manor House Hotel at Moretonhampstead. Journoes also got more off-road driving, but of a mild nature. For those into serious off-roading, Land Rover laid on a driving day a few weeks later at Eastnor where any doubts about Discovery's prowess in the worst conditions were blown apart.

Day and night

The third site at Plymouth was a hive of activity, but not normally seen by outsiders, other than nosey types from LRO. Behind any event are the unsung heroes who make it happen, and this was their base. Part of the Plymouth bus garage had been turned into a flowline where dirty and sometimes battered vehicles entered one end and 100 yards later emerged as sparkling showroom models for the next day's guests.

This work continued day and night on the launch of 80 Discoveries and 45 3.9 and 2.5 litre 1990 model Range Rovers. The latter were used by European guests who had not previously had opportunity to drive the new models. The mini-production line was staffed mainly by Land Rover employees from Solihull, and employment opportunities were given to about twenty local men.

It's always sad to be homeward bound after an enjoyable stay away, so Land Rover arranged one final visit for the dealers. On their way back to the airport, their coaches turned into Buckfastleigh Station which is on the private Dart Valley Railway. A very nice place to be. Steam trains in the heart of the Devon countryside. But what followed was unexpected.

"The train now arriving," turned out to be a Discovery Special. Into the staion came a five-coach train hauled by a Discovery 200 Tdi running on the rails.

Bruff Rail Ltd. makes a rail guidance system which allows a Land Rover to operate as a normal road vehicle or as a rail car. The transition is easily achieved with a lever which sets of bogies on the rails front and rear. Traction is by the normal road wheels which also run on the rails. This system was easily adapted for Discovery, and a tow bar was added to join to the railway carriage coupling.

The purpose of the demonstration was to emphasize the superior torque of the new 2.5 litre turbo diesel engine (195lb ft at 1800rpm). The five coaches weighed over 170 tons, yet the Discovery was able to start off easily. With the clutch open the revs were kept at about 2200, and as the clutch took up, the revs fell back to the figure for maximum torque at 1800rpm. As a safety measure the guard's van was manned so that the trains' brake could be wound on.

A fitting finale to another successful presentation. And feedback from dealers and the press seems unanimous in its approval of the product and its presentation.

Bon voyage, Discovery.

16

"The train now arriving at platform one..."

Bottom right: The little boy who never grew up. Don Green, LR's senior presentations team driver, playing at steam trains.

Washing the vehicles was a round-the-clock task at Plymouth.

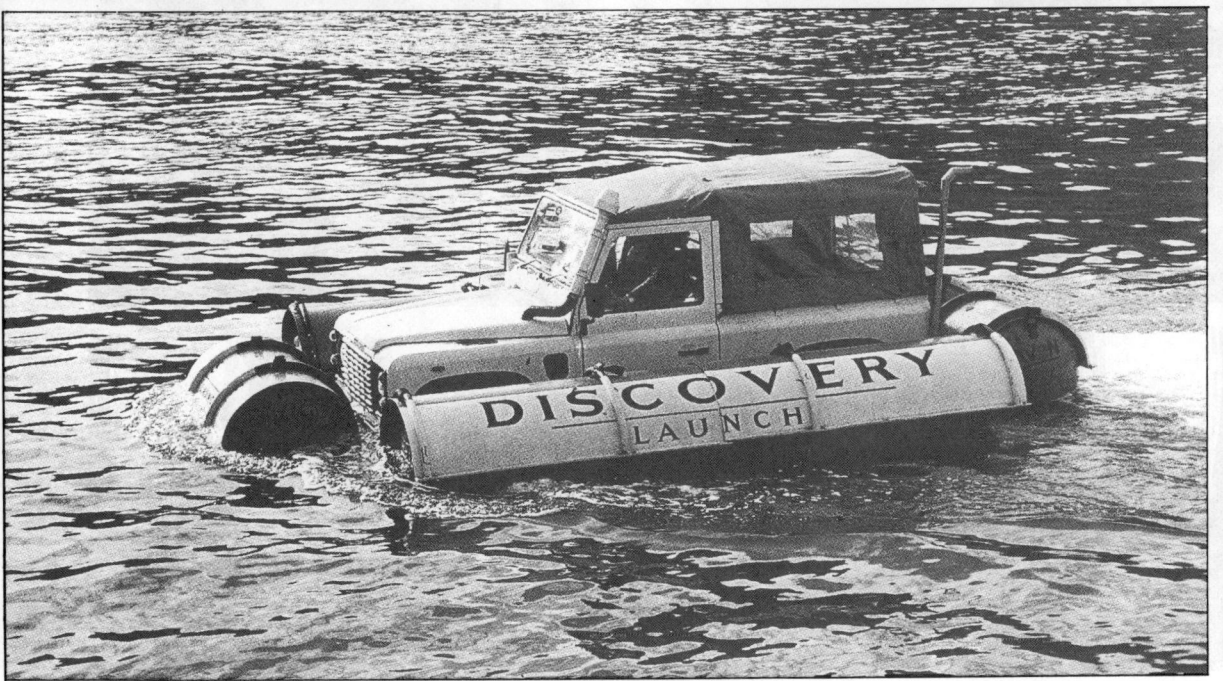

DISCOVERY LAUNCH

Discovery shows Japan the way

On sale from tomorrow, Land Rover's £15,750 Discovery takes on Mitsubishi and Isuzu in the four-wheel drive off-road game. Howard Lees assesses its chances

A PEAK-TIME TV AD WILL tonight introduce the uninitiated to Land Rover's new off roader, the Discovery. Aimed squarely at the Japanese four-wheel drives that dominate the market's middle ground and of crucial importance to Land Rover, Discovery goes on sale tomorrow in V8 petrol and Tdi turbo diesel versions at £15,750. Here we offer driving impressions of the V8, and on pages 38 to 45 we pit the turbo diesel against its two strongest rivals.

The commercial shows a fur-clad team on dog sleds, snowmobiles and ultimately a helicopter chasing a set of tyre tracks across the top of a snow-capped mountain range. As the helicopter blows the snow away to reveal a Discovery parked on the peak, the lead explorer removes goggles and facemask to expose a distinctly oriental visage, and rubs the ice away from the Land Rover bonnet badge. The sign-off line is: "Discovery. It already has its followers."

Produced by BSB Dorland, the advert and its associated air time represents the biggest advertising spend in Land Rover's history.

Questioned as to the production budget for the shoot, Rover Group marketing director John Russell would only describe it as 'very good value'. If it helps Discovery take a healthy slice of Europe's fastest-growing market — one that has never seen a Land Rover product — then it probably is.

Across Europe, three-door off-roaders take the lion's share of the recreational 4wd market and Discovery competes with the short wheelbase Mitsubishi Shogun, Toyota Land Cruiser, Isuzu Trooper and Daihatsu Fourtrak. In the UK, the market is rather different. Here the five-door, long wheelbase Shogun and Trooper dominate the market and initially Discovery must compete head-on in three-door guise. Turbo diesel power accounts for the vast majority of sales — Land Rover expects only 20 per cent of production to be the 3.5 litre V8 — and on price the £15,750 Discovery Tdi comfortably undercuts the market-leading £17,859 Shogun Turbo Diesel.

Discovery also goes on sale in Italy tomorrow, reaching France on December 15 and the rest of Europe by the first quarter of next year. With no fuel injection version of the petrol V8 at launch, and hence no possibility of a controlled catalytic converter, Germany will get only the turbo diesel initially. Rover Group manufacturing planning director Mike Donovan confirmed that Switzerland would get the V8 Discovery by the end of March next year, with a three-way catalyst fuel-injected model to meet strict Swiss emission legislation. For the carburettored V8, a £325 dealer-fitted oxidation catalyst is already available.

Though Land Rover is keeping characteristically tight-lipped about it, a five-door Discovery will also appear next year. Rover Group managing director George Simpson explains what this will mean for Range Rover: "The marketing of more upmarket versions of Discovery against bottom-end Range Rover models is a delicate proposition," he says, confirming that base model Range Rovers will gradually be phased out and the range taken even more upmarket.

To gain a foothold in its intended market in Europe, Discovery needs to be keenly priced. Make no mistake,

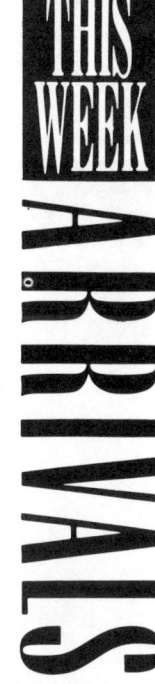

THIS WEEK ARRIVALS

Discovery can be transformed from full 7-seater to capacious load-carrier. Three-door form hampers access to rear seats. V8 shares dual-range transfer box and locking centre diff with Tdi

£15,750 is a bargain basement price that Land Rover has only been able to achieve by launching with its own engines and as a three-door. But though prices are sure to rise as better-equipped and more upmarket versions appear, commercial director Chris Woodwark is adamant that Discovery is already profitable: "Discovery is a good, profitable programme for us." Simpson reinforces this: "The Rover Group is not going to be doing any business that isn't profitable."

In the metal, Discovery has been well described in these pages already. With what is essentially a Range Rover chassis, complete with coil springs and permanent four-wheel drive, Discovery brings a level of 4wd technology to the class that has not been seen before. Combined with a fresh interior approach that matches the Japanese on space and vesatility yet is more recognisably car-like than any of its competitors, Land Rover is not relying solely on its reputation to sell the product.

motorfair STAR COMPARISON

Discovery-v-Sh

Historically, Land Rover always led the field. But then came Mitsubishi's Shogun and Isuzu's
with Discovery on sale from tomorrow. Howard Lees takes all three up to and over the Welsh

FOR THE FIRST TIME IN ITS 40-year history, Land Rover is tackling its competitors on their home ground. With competent, well-built and versatile off-roaders more comfortable than a Land Rover yet without the luxury-segment pricing of the Range Rover, the Japanese have made the recreational four-wheel-drive market their own. Discovery, on sale from tomorrow, is Land Rover's answer.

In the UK this market is dominated by the Mitsubishi Shogun and the Isuzu Trooper. The success of the Shogun has demonstrated that, in the UK at least, budget pricing isn't the most important thing. So, competitively priced though the Discovery is, it will succeed or fail on its merits.

To find out if its on-paper promise is reflected in the metal, we pitted a Discovery Tdi against the best-selling Shogun five-door Turbo Diesel, in its latest intercooled guise, and its nearest rival, the Isuzu Trooper TD. After 1000 miles of road driving and two days of serious off-roading in the Welsh mountains we had our answer.

At £15,750, the Discovery Tdi sits squarely between the £15,299 Trooper and £17,859 Shogun. Most vehicles of this type are highly specified and it is not unusual to spend close to £20,000 putting one on the road. With the optional electric pack (central locking, head-lamp washers, electric front windows and mirrors), plus a remote alarm, twin sunroofs, foldaway rear seats, front and side protector bars, towbar and metallic paint, our test Discovery weighed-in at £18,377. A limited slip diff and headlamp washers in addition to the standard-fit electric windows and central locking take the Trooper Duty to £16,198, while the air-conditioned Trooper Citation we tested is listed at £17,999. At the top of the league is the Shogun — its £1250 Diamond option pack (limited slip rear diff, electric windows and sunroof), plus protector bars and towbar lift the tag to £20,312.

All three share iron-block, alloy-head, four cylinder turbo diesel engines and five-speed gearboxes. Top of the class for power and torque is Land Rover's new 2.5-litre intercooled direct injection 200 Tdi, its 111bhp at 4000rpm and 195lb ft of torque at 1800rpm comfortably shading the intercooled but indirect injection 2.5-litre Mitsubishi unit's 95bhp at 4200rpm and 173lb ft of torque at 2000rpm. While it matches the Mitsubishi's 95bhp at only 3890rpm, Isuzu's 2.8-litre direct injection unit lacks an intercooler and the consequent lower boost pressure yields a modest 153lb ft of torque at 2100rpm.

The Discovery's drivetrain also stands out from the crowd. The Trooper and Shogun share a typical Japanese dual range transfer box, part-time four-wheel drive layout ▶

gun-v-Trooper

rooper, making off-road four-wheel-drive fun Japanese territory. Now comes the fightback, ountains to see whether Land Rover is back in the vanguard. Photography by Stan Papior

with freewheeling hubs and limited slip rear differentials. The Discovery sticks with Land Rover's familiar dual range transfer box and lockable centre differential to provide permanent 4wd. And while all three have substantial steel ladder chassis, Land Rover's suspension layout is very different from the Japanese way. Like the Range Rover, the Discovery has coil springs and live axles — both Mitsubishi and Isuzu favour torsion bar-spring independent wishbone front suspension and a leaf-sprung live axle at the rear, with anti-roll bars at front and rear.

The Discovery and the Shogun share 205 R16 tyres, Goodyear Wranglers on the Land Rover and Dunlop SP TGR on the Mitsubishi, while the Trooper's 15ins rims are shod with 235/75 section BF Goodrich All-Terrain rubber. All have front disc brakes, ventilated on both Trooper and Shogun, while at the rear the Shogun's drums are at odds with the others' discs. All three have power-assisted recirculating ball steering as standard.

With the exception of the roof, the Discovery's body panels are aluminium alloy rather than steel pressings. Their light weight doesn't help on the weighbridge — as tested the Tdi's kerb weight of 4432lb is substantially more than the 4214lb Shogun and 4002lb Trooper. And the Discovery is only a three-door.

PERFORMANCE AND ECONOMY
A clear win for the Discovery here: 111bhp powers it to a class-leading top speed of 92mph, well ahead of the 85mph Shogun and 86mph Trooper. Its extra weight tempers the Tdi's advantage under acceleration, but 0-60mph in 17.1secs is 1.7secs up on the Shogun and

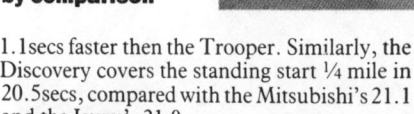

Discovery interior (left and right) is refreshingly different. Shogun (below) shows its commercial origins. Trooper (bottom) looks tacky by comparison

1.1secs faster then the Trooper. Similarly, the Discovery covers the standing start ¼ mile in 20.5secs, compared with the Mitsubishi's 21.1 and the Isuzu's 21.0.

Both the Discovery and Shogun engines have a broad spread of power, with decent torque available from 2000rpm to within a couple of hundred rpm of their smoothly governed 4500rpm maximum. The Tdi engine is more responsive and will pull more strongly from low rpm, but both put the Trooper's 2.8 to shame. This is the least responsive and has easily the narrowest usable power band, coming on boost well past 2000rpm and running out of steam above 4000rpm. It is reluctant to reach its governed peak of 4250rpm, and also suffers from a switch-like throttle with precious little travel.

These impressions are reinforced by the acceleration figures in each gear. The Discovery's longer legs, 25.1mph per 1000rpm in top against the Shogun's 20.8, give the Mitsubishi the advantage in fourth from 30-50mph — either will still walk away from the high-geared Trooper.

None of the contestants is without fault in the

Discovery's coil springs and live axles give most ground clearance and best articulation. Easily most capable and comfortable off-road

gearchange department. Land Rover's five-speeder gets increased width gears and a new shift gate for 1990, but the shift can be a bit awkward and first is still too close to reverse for comfort. With its new, stronger gearbox the Shogun has lost the slick, precise gearchange it used to have and become altogether stiffer and notchier. The Trooper's change is the lightest, but a long lever imparts a sloppy feel.

During the comparison, including the off-roading stages, the Discovery returned an average of 24.9mpg. The Trooper's larger DI engine returned 23.1mpg while the Shogun's indirect injection diesel managed just 19.8mpg, these figures demonstrating the efficiency of direct injection. Over the full road test period, including performance testing, the pattern was repeated — 23.9mpg for the Discovery, 22.1 for the Trooper and 20.2 for the Shogun. All three have usefully large fuel tanks: 18.2gals for the Trooper, 19.5 for the Discovery and 20.2 for the Shogun.

RIDE AND HANDLING

The long travel suspension and high centre of gravity that characterise a big off-roader bring the inevitable penalties in outright handling. Mitsubishi's Shogun has a deservedly good reputation as a road vehicle, and its 3.5 turns lock to lock steering is both the highest geared and most positive. At 3.8 turns the Trooper is lower geared, but has a weightier feel to it that provides more communication with the front wheels — what lets it down is rather too much feedback on an uneven road.

Not surprisingly, given the similarity of its chassis, the Discovery's steering feels like a Range Rover's. Gearing is similar to the Trooper's and the level of power assistance is on a par with the Shogun, but the damper fitted to tame bump steer off-road imparts a vague, lifeless feel at speed.

With no anti-roll bars and long travel coils at each corner, the Discovery rolls a good deal when cornering hard on the road. But with the fine traction and balance provided by its permanent 4wd, the Discovery has significantly more outright grip than the Shogun. On the limit, its Goodyear Wranglers slide progressively into an understeering drift, yet backing off the throttle does no more than pull the nose back into line.

The Shogun's anti-roll bars keep roll better checked, though with the rear wheels only being driven it ploughs into understeer earlier and its Dunlop SP TGRs work harder to deliver less outright grip. Lift the throttle on the limit and the tail can get out of shape, especially in the wet. Bottom of the list is the Trooper, which despite its wider 235 section BF Goodrich All-Terrains has the least grip, wet or dry, and gets out of shape earlier than the Mitsubishi.

All three have decent brakes with plenty of stopping power, but the Discovery offers better feel and pedal progression and is ultimately able to generate more stopping power than the other two — a full 1.0g at lock-up compared with the Shogun's 0.68g and the Trooper's 0.7g.

It may not have the suppleness of the Range Rover, but Discovery's coil springs still provide a ride superior to either of the Japanese off-roaders. Despite its leaf rear springs, the Shogun puts up a creditable showing, helped by its independent front suspension, but it can't match the Discovery over smaller bumps. Though the Trooper shares its suspension ▶

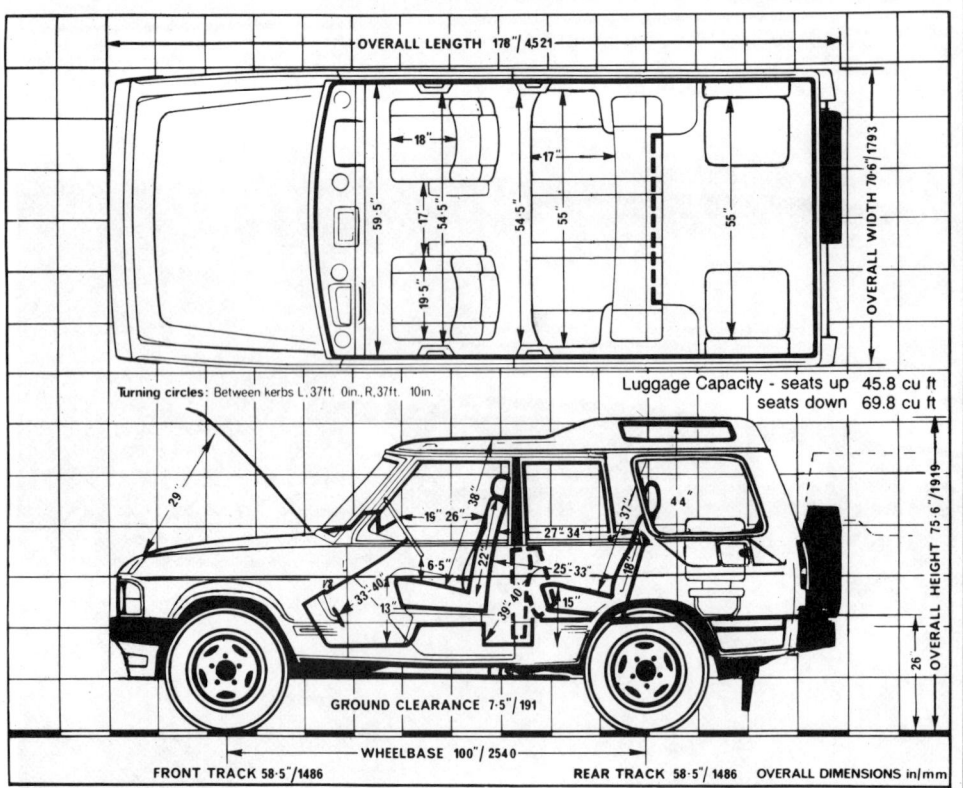

PERFORMANCE

MAXIMUM SPEEDS

Gear	mph	km/h	rpm
Top (Mean)	92	148	3650
(Best)	92	148	3650
4th	84	135	4350
3rd	63	101	4500
2nd	41	66	4500
1st	23	38	4500

ACCELERATION FROM REST

True mph	Time (secs)	Speedo mph
30	4.7	32
40	7.6	43
50	11.5	54
60	17.1	65
70	25.1	75
80	37.1	86

Standing ¼-mile: 20.5secs, 63mph
Standing km: 38.8secs, 81mph
30-70mph thro' gears: 20.6secs

ACCELERATION IN EACH GEAR

mph	Top	4th	3rd	2nd
10-30	—	15.4	8.8	5.0
20-40	20.8	11.3	6.8	5.2
30-50	16.1	9.7	7.4	—
40-60	16.7	11.0	9.7	—
50-70	20.9	14.3	—	—
60-80	32.4	21.3	—	—

FUEL CONSUMPTION
Overall mpg: 23.9 (11.8 litres/100km)
Touring mpg*: 32.3mpg (8.7 litres/100km)
Govt tests mpg: 30.5mpg (urban)
42.3mpg (steady 56mph)
28.9mpg (steady 75mph)
Grade of fuel: Diesel
Tank capacity: 19.5 galls (89 litres)
Max range*: 630 miles

* Based on Government fuel economy figures: 50 per cent of urban cycle, 25 per cent each of 56/75mph consumptions.

BRAKING
Fade (from 63mph in neutral)
Pedal load (lb) for 0.5g stops

	start/end		start/end
1	20-10	6	30-55
2	22-10	7	35-65
3	20-25	8	30-60
4	20-35	9	30-50
5	25-47	10	30-38

Response (from 30mph in neutral)

Load	g	Distance
10lb	0.15	201ft
20lb	0.38	79ft
30lb	0.60	50ft
40lb	0.88	34ft
50lb	1.0	30ft

WEIGHT
Kerb 4432lb/2012kg
Distribution % F/R 51/49
Test 4795lb/2177kg
Max payload 1559lb/708kg
Max towing weight 8800lb/4000kg

TEST CONDITIONS
Wind 7mph
Temperature 16deg C (61deg F)
Barometer 1012mbar
Surface dry asphalt/concrete
Test distance 1100 miles

Figures taken at 4100 miles by our own staff at the Lotus Group proving ground, Millbrook.

All *Autocar & Motor* test results are subject to world copyright and may not be reproduced without the Editor's written permission.

LAND ROVER DISCOVERY TDi

ENGINE
Longitudinal, front, permanent four wheel drive. Capacity 2495cc, 4 cylinders in line.
Bore 90.5mm, **stroke** 97.0mm.
Compression ratio 19.5 to 1.
Head/block al alloy/cast iron.
Valve gear ohv, 2 valves per cylinder.
Ignition and fuel direct diesel injection, Garrett T25 turbocharger.
Max power 111bhp (PS-DIN) (83kW ISO) at 4000rpm. **Max torque** 195lb ft (265 Nm) at 1800rpm.

TRANSMISSION
5-speed manual.

Gear	Ratio	mph/1000rpm
Top	0.770	25.1
4th	1.000	19.3
3rd	1.397	13.9
2nd	2.132	9.1
1st	3.692	5.2

Final drive ratio 3.538 to 1. Transfer ratio: high 1.222, low 3.320, lockable centre differential.

SUSPENSION
Front, live axle, radius arms, Panhard rod, coil springs, telescopic dampers.
Rear, live axle, radius arms, upper 'A' frame, coil springs, telescopic dampers.

STEERING
Recirculating ball, power assisted, 3.8 turns lock to lock.

BRAKES
Front 11.8ins (299mm) dia discs.
Rear 11.4ins (290mm) dia discs.

WHEELS/TYRES
Pressed steel, 7ins rims. 205 R16 Goodyear Wrangler.

PRODUCED AND SOLD IN THE UK BY
Land Rover Ltd
Lode Lane,
Solihull, W. Mids B92 8NW.
Tel: 021-722 2424

COSTS

Prices
Total (in UK)	£15,750
Delivery, road tax, plates	£375
On the road price	£16,125
Options fitted to test car:	
Front bull bars & lights	£181
Side protectors	£150
Side stripes	£44
Electric pack	£525
2nd sunroof	£325
7-seat option	£375
Alarm	£282
Metallic paint	£250
Towbar	£120
Total as tested	£18,377

SERVICE
Major service 24,000 miles — service time 4.95 hrs. Intermediate service 12,000 miles — service time 4.25 hrs. Oil change 6000 miles — service time 0.30 hrs.

PARTS COST (inc VAT)
Oil filter	£10.49
Air filter	£18.40
Brake pads (2 wheels) front	£50.01
Brake pads (2 wheels) rear	£56.32
Exhaust complete	£459.71
Tyre — each (typical)	£68.77
Windscreen	£119.60
Headlamp unit	£24.50
Front wing	£111.55
Rear bumper	£80.50

WARRANTY
12 months/unlimited mileage, 3 years anti-corrosion, 12 months breakdown recovery

EQUIPMENT

Air conditioning	£1290
Alloy wheels	£TBA
Power assisted steering	●
Steering rake adjustment	●
Headlamp wash	E
Seat tilt adjustment	—
Lumbar adjustment	—
Split rear seat	●
Remote boot/hatch release	●
Internal mirror adjustment	E
Flick wipe	●
Programmed wash/wipe	●
Revcounter	●
Lockable glovebox	●
Radio/cassette player	●
Electric aerial	●
4 speakers	●
Electric windows (front)	E
Central locking	E
Tailgate wash/wipe	●
Tinted glass	●
Sunroof	●
Metallic paint	£250

E £525 option pack ● Standard — Not available

1 Indicator plus dip, **2** Heated rear screen, **3** Rev counter, **4** Warning lights, **5** Speedo, **6** Rear wash/wipe, **7** Wipers, **8** Lighter, **9** Heater, **10** Clock.

◀ layout with the Shogun, it offers a coarse, rather jiggly ride.

Off-road, the pedigree of Discovery's chassis and drivetrain outclasses the Japanese by a large margin. The ride is superb — long travel, supple and well-controlled wheel movement with fine articulation keeping all four wheels on the ground for more of the time, providing usefully superior grip. A lower ratio transfer box allows better engine braking down steep descents and combines with the flexible engine to allow most slopes to be climbed in third gear. There's more ground clearance at the front because of the live axle, and that steering damper comes into its own by preventing any potentially thumb-wrenching kickback through the steering wheel. All in all, a convincing win, on or off the road, for the British newcomer.

AT THE WHEEL

Despite its size, the Trooper has only just enough rearward seat travel for 6ft-plus drivers, while the Discovery has room to spare. The Shogun is somewhere between the two. All three have generous headroom. Like the Isuzu, Mitsubishi's offering has a rake-adjustable steering wheel. Its driving position also benefits from adjustable side bolsters on the seat back that give it the best lateral support, but the Discovery runs it close and actually has the most comfortable chairs.

Well thought out and put together though the Shogun's dashboard is, it still displays commercial vehicle rather than car origins. Discovery brings fresh thinking to the interior, with clear instruments and well-placed switchgear and minor controls. With two sunroofs and the 'alpine' windows in the raised roof section, it is airy and spacious inside, with the rearmost seats folding to below window level and thus not restricting rear three-quarter visibility as they do in the Shogun. Again the cheap-feeling Trooper is a poor third.

Only the Trooper Citation has air conditioning as standard, and both the Japanese 4wds have simpler heater controls than the Discovery's five-slider, hard-to-reach carry-over from the Range Rover. Neither the Trooper nor Discovery are over-endowed with fresh-air ventilation compared to the Shogun's impressive throughput.

COMFORT AND SPACE

These are big, bluff shapes to force through the air, and at speed none of the cars is particularly free of wind noise. Worst offender is the Trooper, with the Discovery and Shogun only suffering above 85mph. Direct injection does cause problems with combustion noise, and both the Trooper and Discovery have a noticeable diesel growl at speed. Smoothest and quietest engine by some margin is the Shogun's balancer shaft-equipped 2.5-litre.

Rear seat passengers have the most legroom in the Discovery, closely followed by the Shogun with the Trooper again a poor third. But with only a single door each side, Discovery's rear seat is nowhere near as accessible as those of the five-door Shogun and Trooper.

A pair of forward-facing fold-down seats in the rear make the Mitsubishi the most convincing seven-seater. The fold-away inward-facing seats in the Discovery have lap seat belts only and offer little headroom for tall adults. Only the Land Rover has a split-fold rear seat, although with an overall length 3ins up on Discovery, the Shogun has the longest luggage area. The Trooper's limited front and rear legroom gives it a bigger load area than the Discovery too.

For interior stowage the Discovery wins hands down. As well as bins in all three doors, dash, either side of the rear seat and between the front seats, there are pockets for maps above the sun visor, stowage nets in the stepped roof section at the rear and pouches on the back of the front seats. There's also a zipped bag on the back of the rear seats to carry the removable sunroof, and a non-slip rubber mat allows the top of the dash to be used as a shelf with no danger of things sliding off even during extreme cornering.

FINISH AND EQUIPMENT

As tested, equipment levels in all three off-roaders are similar. All have electric windows, central locking and a modest stereo system. The Trooper Citation, while costing the least, boasts air conditioning that is an expensive option on the other two. Only the Discovery has electric mirrors, but the Diamond option pack-equipped Shogun boasts a huge electric sunroof. ▶

Permanent 4wd is key to Discovery's fine balance. Shogun still capable but Trooper's tyres lack grip. Both rivals have bigger load areas than Land Rover (top)

HOW DISCOVERY, SHOGUN AND TROOPER COMPARE

	DISCOVERY	SHOGUN	TROOPER		DISCOVERY	SHOGUN	TROOPER
PRICES				**ACCELERATION FROM REST (secs/speedo mph)**			
Total in UK	£15,750	£17,895	£17,999	0-30	4.7/32	4.9/32	4.6/32
Total as tested	£18,377	£20,132	£18,249	0-40	7.6/43	8.3/43	7.8/4.2
Length	178ins	181.1ins	176ins	0-50	11.5/54	12.5/54	11.7/53
Width	70.6ins	66.1ins	64.9ins	0-60	17.1/65	18.8/65	18.2/64
Wheelbase	100ins	106.1ins	104.3ins	0-70	25.1/75	28.5/76	28.0/74
Height	75.6ins	74ins	70.9ins	Stand'g ¼m, secs/mph	20.5/63	21.1/63	21.0/64
ENGINE				Stand'g km, secs/mph	38.8/81	39.8/74	39.6/77
Capacity, cc	2495	2477	2771	30-70 thro' gears, secs	20.6	23.6	23.4
Max power, bhp/rpm	111/4000	95/4200	95/3800	**ACCELERATION IN TOP/4th**			
Max torque, lb ft/rpm	195/1800	173/2000	153/2100	20-40	20.8/11.3	11.9/8.7	22.4/14.3
TRANSMISSION Ratios/mph per 1000rpm				30-50	16.1/9.7	10.4/8.5	21.7/11.4
Top	0.77/25.1	0.829/20.8	0.809/24.1	40-60	16.7/11.0	12.5/6.2	21.0/12.0
4th	1.00/19.3	1.000/17.2	1.000/19.5	50-70	20.9/14.3	17.4/16.1	23.2/15.9
3rd	1.397/13.9	1.395/12.3	1.404/13.9	60-80	32.4/21.3	–/–	–/28.1
2nd	2.132/9.1	2.261/7.6	2.314/8.4	**ACCELERATION IN 3rd/2nd mph**			
1st	3.692/5.2	3.918/4.4	3.767/5.2	10-30	8.8/5.0	7.4/4.7	9.4/4.9
Final drive	3.538	4.875	4.30	20-40	6.8/5.2	6.1/–	7.5/–
Transfer ratio high/low	1.222/3.32	1.0/1.925	1.0/1.870	30-50	7.4/–	7.7/–	7.1/–
BRAKES				40-60	9.7/–	–/–	7.1/–
Front	11.8ins discs	10.9ins v. discs	10.1ins v. discs	**FUEL CONSUMPTION (mpg)**			
Rear	11.4ins discs	10ins drums	10.4ins discs	Overall Road Test	23.9	20.2	22.1
WHEELS AND TYRES				Comparison	24.9	19.8	23.1
Material/width	Steel/7ins	Steel/6ins	Steel/6ins	Tank capacity, galls	19.5	20.2	18.2
Tyre size and make	205R16	205R16	235/75R15	**WEIGHTS**			
	Goodyear	Dunlop	BF Goodrich	Kerb weight, lb	4432	4214	4002
	Wrangler	SP TGR	All-Terrain	Max towing weight, lb	8800	7275	6612
PERFORMANCE Max speed, mph/rpm				**EQUIPMENT**			
				Air conditioning	£1290	£1546	●
Top (mean)	92/3650	85/4100	86/3550	Limited slip differential	—	D	●
(best)	92/3650	88/4250	90/3750	Internal mirror adjustment	E	—	●
4th	84/4350	77/4500	83/4250	Electric windows (front)	E	D	●
3rd	63/4500	55/4500	59/4250	Central locking	E	●	●
2nd	41/4500	34/4500	36/4250	Sunroof	●	D	—
1st	23/4500	20/4500	22/4250	Metallic paint	£250	£150	●

D — £1250 Diamond option pack. E — £525 option pack.

Clockwise from left: Discovery's 2.5-litre unit offers 111bhp at 4000rpm; Shogun's 2.5, 95bhp at 4000; Trooper's 2.8, 95bhp at 3890. All have four-pot turbo diesel engines

◀ With alloy panels, the Discovery still has the generous panel gaps that characterise the Range Rover, and its paint finish is not as smooth as the Shogun's. How the test car's light blue interior will stand up to muddy wellies has yet to be seen, though rubber mats and waterproof seat covers are available. The Shogun is solidly engineered and well-built, and first impressions say it is better here than the less durable-feeling Discovery. The Trooper doesn't exude the same air of quality.

VERDICT

Though by most standards an honest-enough vehicle that offers good value, it's the Trooper that's bottom in this comparison. Marginally faster than the Shogun, it hasn't the engine flexibility to cut it on the road. By a small though significant margin it has the least grip, the harshest ride and is the noisiest at speed.

Though starting to look expensive, the Shogun still has a great deal going for it. It is roomy, well-built, has the smoothest engine and is not too far adrift of the Discovery in ride quality. It has the benefit of a three-year unlimited mileage warranty — three times that offered by the competition — and over the past few years has built up a loyal following.

Faster, more economical, better-riding and with the extra traction and balance of permanent 4wd, Discovery has the measure of its rivals. The cleverly designed and well-executed interior is way ahead of the opposition and it has a clear advantage should anyone actually venture off-road. Where Land Rover will struggle initially is in trying to sell the Discovery to people who have grown used to five doors — but its other qualities more than make up for a slight struggle to get into the rear. It has been a long time coming but, with the Discovery, Land Rover shows just how good a 'recreational' off-roader can be. With the right build quality, this new champion of Britain's motor industry is good enough to send the Japanese back to the drawing board. ■

THIS WEEK ARRIVALS

Off-road ride is markedly superior to that of rivals

LAND ROVER DISCOVERY V8

ENGINE
Longitudinal, front engine/permanent four wheel drive.
Capacity 3528cc, eight cylinders in 90deg vee.
Bore 88.9mm, **stroke** 71.1mm
Compression ratio 8.1:1.
Head/block Al alloy/al/alloy
Valvegear Ohv, 2 valves per cylinder.
Fuel and ignition Twin SU carburettors, electronic ignition.
Max Power 145bhp (PS Din) (108kW ISO) at 5000rpm.
Max torque 192lb ft (260Nm) at 2800rpm.

TRANSMISSION
5-speed manual.
Ratios: Top 0.77, 4th 1.00, 3rd 1.397, 2nd 2.132, 1st 3.321.
Final drive ratio 3.538. **Transfer ratio:** high 1.222, low 3.320, lockable centre differential. 25.1mph/1000rpm in top.

SUSPENSION
Front, live axle, radius arms, panhard rod, coil springs, telescopic dampers.
Rear, live axle, radius arms, upper 'A' frame, coil springs, telescopic dampers.

STEERING
Recirculating ball 3.8 turns lock to lock, power assisted.

BRAKES
Front, 11.8ins (298mm) dia discs.
Rear, 11.4ins (290mm) dia discs.

WHEELS/TYRES
Pressed steel, 7ins rims, 205 R16 Goodyear Wrangler tyres.

DIMENSIONS
Length 178ins (4521mm)
Width 70.6ins (1793mm)
Height 75.6ins (1919mm)
Wheelbase 100ins (2540mm)
Weight 4359lb (1932kg)

PERFORMANCE (claimed)
0-60mph 12.7secs. Maximum speed 102mph.

FUEL CONSUMPTION
13.0 urban, 26.2 constant 56mph, 19.6 constant 75mph.

PRICE
Total (in UK) £15,750.

DRIVING THE V8

WITH ITS TWIN SU carburettored V8 burbling away under the bonnet, Discovery sounds like a Range Rover used to before the latest 3.9 litre Vogue came along. The venerable all-alloy pushrod unit lacks mechanical sophistication and drinks fuel at an alarming rate, but 145bhp at 5000rpm and a hefty 192lb ft of torque at 2800rpm are claimed to make Discovery the quickest off-roader in its class — Land Rover quotes a top speed of 102mph and 0-60 in 12.7secs.

With a seamless power delivery and impressive torque from as little as 2000rpm, the 3.5 litre V8 can hustle the 4400lb Discovery along with verve. Generally smooth and quiet, the V8 runs out of steam and starts to get a little harsher beyond 5000rpm; like the Range Rover, there is no red line or rev limiter but very little to gain by exceeding 5500rpm. The five-speed gearbox still needs familiarity and care to get the best out of it, but though it uses the old Range Rover's gear-driven transfer box most of the characteristic whine on the over-run is mercifully exorcised.

Not surprisingly, the handling has a distinctly Range Rover-like feel, with no anti-roll bars to check the considerable body lean angles but decent grip and predictable understeer on the limit. Permanent four-wheel drive gives it the traction edge on tricky surfaces, as well as preventing Discovery getting out of shape if provoked; lifting the throttle on a tightening, off-camber bend will do little more than quell the understeer and bring the nose back into line. On the downside, with a hefty steering damper Discovery inherits the Range Rover's vague, lifeless steering.

Inside, only a manual choke knob distinguishes the V8 from the turbo diesel Tdi, which means it has the same roomy interior, well laid-out controls and generous stowage that set it apart from the crowd. Our test V8 had the optional high-spec stereo, which doesn't offer a £450 improvement in sound quality even though it provides remote controls for the volume and tuning on the edges of the instrument binnacle.

Off-road, the V8 is extremely capable — easy and forgiving to drive, with plenty of grip from its 205 R16 Goodyear Wrangler tyres. There's not as much engine braking available as on the turbo diesel, but the very low ratio available in the dual-range transfer box offers plenty in first gear for even the most fearsome descent.

Fine articulation and good ground clearance make the V8 almost unstoppable with the centre diff locked. The V8 Discovery not only out-performs the opposition but delivers a markedly superior ride both on and off-road.

Land Rover chose image-maker Conran Design for the interior of the new Discovery and the results are truly striking. Howard Lees talks to the men whose work up to now has been more familiar in the high street than to the 4wd enthusiast

INSIDE STORY

Lifestyle board identifies market as 'young Eurotech', all Braun razors and BMWs

MENTION THE ADVENT OF HIGH street style in the UK and the Conran name — of Habitat, Heals and Mothercare fame — comes instantly to mind. So, in the search for an interior that would reflect the values of the new Discovery's maximum potential market, Land Rover called on Conran Design.

The studio's involvement with project Jay — Land Rover's code name for Discovery — dates back to August 1986. Along with another non-automotive design consultancy, a motor industry design group and Land Rover's own styling studio, Conran was commissioned to produce a complete interior concept for the new off-roader. Mike Donovan, now Rover Group's manufacturing planning director but then in charge of project Jay at Land Rover, wanted all the interiors defined before making the final choice.

Donovan briefed Conran Design thoroughly on what he wanted to achieve and on the market at which Jay was aimed. Howard Chapman, the designer put in charge of the project, explained: "To some extent it's true that Land Rover came to us with a customer profile. Land Rover wanted a lifestyle vehicle, but part of the design exercise was to identify exactly which type of lifestyle."

The first stage was to produce half a dozen or so 'lifestyle boards' — collages of various objects, colours and textures likely to play a part in the life of various types of potential customer. From the start, Conran was asked to look at both three and five-door versions and all the full-size modelling was done as a five-door. Land Rover has still not officially admitted that a five-door Discovery exists and Tony Rowe, Conran Design director responsible for the project, will say little about it: "Let's just say that we ended up knowing what a five-door would look like."

The final lifestyle model for the three-door was a group summed up as 'young Eurotech' — all Braun razors and BMWs with a predominance of blue, black and charcoal grey. Chapman confirmed that a five-door study was angled more at the family through a group

Mock-up of facia (below) shows how closely Discovery (below left) sticks to Conran Design specifications

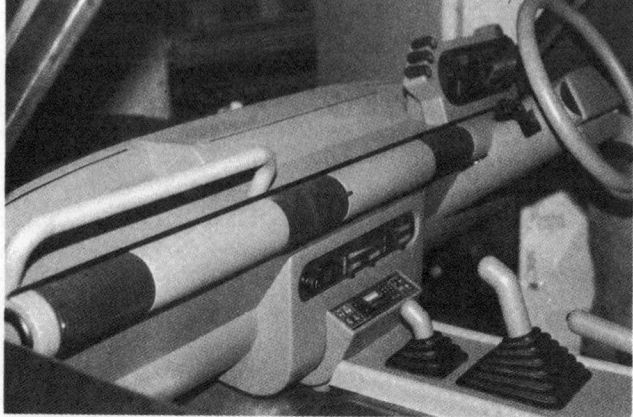

tagged as 'country casual' — Barbour jackets and Hunter wellies.

"Once we had established the lifestyles we were aiming at, we could start the creative flow." That meant producing a series of initial sketches of details such as the doors, dash and steering wheel as well as the whole interior. Chapman was keen to make all aspects of the interior integrate: "The intention was to carry an integral theme throughout the interior, rather than simply spending all the available money around the driver."

Rowe explained the design process: "In simple terms, on the one hand you have the functioning of the vehicle and on the other the image or appearance of the vehicle. One part of the exercise was to look at the function, in terms of storage, comfort, readability of the instruments and so on. The other aspect is how those components manifest themselves in a visual sense — what image that vehicle gives to a person who gets into it.

"Firstly we tried to pinpoint a number of key images by a process we call orientation. We might use the clothing these people wear, the products they buy, the sort of houses they live in and the holidays they take to get a feel for the aspirations of that person. Having established this image as one element, we then marry this to the broad functional requirements of the design. The two requirements tend to leapfrog each other until finally we produce a vehicle that has the appropriate functions and features but also the right image for this lifestyle."

Of course, Conran had to work within the existing vehicle parameters. Defining the dimensions of the interior was a full-size bodyshell that was delivered to Conran's central London studios within a month of the start of the project. Craned in through a third floor window, it was actually a five-door Range Rover; dimensionally the two are very close and some components such as the main dash structure and heater unit are shared.

That caused some problems for Chapman: "The heater was the main stumbling block — it is massive. But we developed a shelf-like aspect to the dashboard that let the heater protrude from it, which Land Rover was very keen on." Once the dash concept had been developed, it was tied in with ideas for the door trims and remainder of the interior, and very early on the sketches began to resemble the interior as it is now. Even the light blue colour scheme and 'golfball' dimpled texture showed up clearly very early on — Chapman wanting nothing to do with the grained leather finish almost universally used for dash and door trims elsewhere.

Land Rover was closely involved as the project developed, with Donovan's team paying at least monthly and sometimes bi-weekly visits to the studio. By now six designers and another half a dozen model makers were working on the Jay project. Conran Design was also working with JVB, a feasibility company brought in by Land Rover to examine design from the engineering,

Designer Howard Chapman: 'Huge heater the main stumbling block'

production, safety and homologation point of view, while Land Rover's component supplier was involved early on.

Both Chapman and Rowe visited Land Rover at Solihull a number of times, and on one occasion Chapman was bounced around the jungle track in a Land Rover with demonstration guru Don Green at the wheel: "I was terrified, but from holding onto the back of his seat I realised that we needed grab handles for the rear passengers as well as the front. It was also clear that we really didn't need much width at the top of the front seats, and were able to develop the grab handles integral with the seat backrests."

The detailed sketches progressed through scale models to a full-size interior built inside the Range Rover shell. By March 1987, the contest had narrowed down to a straight fight between the Conran interior and a Land Rover modified version, but right from the beginning Chapman was aware that Conran Design wasn't the only company working on the design: "It was interesting because we were told at the start that we were in competition — that made the first stages crucial."

All the companies involved were paid for the work they did, but even when the Conran proposal was selected, Chapman's work was far from over. Working with Land Rover as it produced its own full-size model at Solihull, the basic design project finished 18 months ago 'when Land Rover felt it could take over.' Since then Chapman has developed many accessory ideas designed to integrate with the centre console and rear luggage area, to the point where Land Rover has to decide whether to produce them; the neat detachable soft bag between the front seats is one example of a host of concepts that include a cool box, phone and computer console and camera bag.

Despite Conran's limited previous experience in automotive design, the contract held no terrors for Rowe: "The design of a vehicle interior is a very complex job, but the elements are straightforward once you understand ergonomics, safety, tooling, the principles of assembly, maintenance and testing. We have designed many, many different products — so, yes, a vehicle is complex in its totality, but as long as you approach the job in a logical manner, it is relatively straightforward."

Straightforward or not, Discovery's interior is a striking yet functional design that lifts it way above the van-like insides of its Japanese competition. Choosing the Conran proposal must have been a tremendous gamble for a company like Land Rover, but it looks as though it will pay off. ■

Project director Tony Rowe: 'Establish the image and marry it with functional design'

Early sketch of fold-up boot seats (below) shows basic pattern. Shapes stayed constant to production (below left)

The Discovery has a couple of rather hard acts to follow. Other manufacturers have tried to copy the original Land Rover concept – and none has ever really succeeded. And the Range Rover has carved a special niche entirely to itself, with sales following its increasingly upmarket image. Never slow in grasping a marketing opportunity, the Japanese moved in to fill the gap between the two sides of the Land Rover family, with a whole range of four-wheel drive 'leisure' vehicles, with a tough exterior image combined with saloon-car like interior comfort. But while their cross country ability was in most cases good, none has ever managed to match the sheer unstoppability of either the still-rugged Land Rover or even the latest leather-bound, walnut-veneered 3.9-litre Range Rover Vogue SE.

But Land Rover was sitting on one of the best-kept secrets in the British motor industry. Of course everyone *knew* that there was a new, small Range Rover on the drawing board. They were in for a surprise when the Discovery went public. Here was a 4wd vehicle aimed to slot into the space which Land Rover had deliberately made for itself in its model range. And what had been doing the job before? For a long time Mitsubishi had this £16,000 to £18,000 slice of the 4wd market more or less to itself, with Isuzu a comparative newcomer with the Trooper. But the big shock came when the Discovery's price was revealed – both the 3.5 V8 and 2.5 turbodiesel version wearing a £15,750 price tag.

With the size and power outputs of the three petrol engined rivals varying so widely, we decided that it would be fairer to test the top turbodiesel versions, which will probably account for the majority of Discovery sales. The 'target' car had to be the Mitsubishi Shogun, which has a 25 per cent share of this part of the 4wd market.

The Discovery has been deliberately badged as a Land Rover. But the Range Rover family likeness is obvious – and it is more than just skin deep. Underneath there is the same 100in wheelbase chassis, with coil sprung live axles front and rear, with permanent four-wheel drive. Currently the Discovery is only made in three-door form, although a five-door version is on the stocks. Even so, it is close to three inches longer and, because of the high roof, over five inches taller. The inner body panels are steel, but the outer ones are mainly in corrosion-resistant aluminium apart from the big single-piece steel roof pressing. The turbodiesel, direct injection 2.5 litre engine has been developed specially for the 200 Tdi Discovery, developing 111bhp at 4000rpm. But the amount of

VOYAGE OF

The wraps are finally off the 'secret' Land Rover. The Discovery is pitched between the rugged Land Rover and styl

Discovery is Land Rover's answer to imported 4wd leisure vehicles like the popular Shogun and Trooper. The beefy newcomer slots neatly into a niche below Range Rover — and undercuts the Japanese rivals on price

torque is truly impressive, with 195lb ft coming in at only 1800rpm.

The all-wheel drive transmission is 'Mk I' Range Rover, with the original gear drive to the two-speed transfer box being used, rather than the newer and quieter chain system on the latest models. When the going gets really sticky the centre diff can be locked.

Isuzu's Trooper is a couple of inches shorter than the Discovery, with a squared-up, chunky five-door body in its long wheelbase form. As with the Shogun, normal drive is to the rear wheels, with front-wheel drive being added automatically through locking front hubs when the selector is moved into either high or low ratio. Put the selector into high ratio two-wheel drive and reverse a metre or so, and the hubs automatically unlock. The Trooper's 2.8 direct injection diesel does not have an intercooler in its turbo system. Power output is 95bhp, while maximum torque of 153lb ft is developed at 2100rpm.

Like the Trooper, the LWB Mitsubishi Shogun has independent front suspension, and a live axle at the rear, with semi-elliptic springs. The drivetrain too is similar, with automatic locking front hubs. Neither of the Japanese cars has locking diffs. The 2.5 engine has the more usual indirect injection system, but the twin counter-rotating balancer shafts damp out any roughness. The intercooler is located horizontally over the engine, with air drawn in through a bonnet-top scoop. Output is 94bhp at 4200rpm, with the 173lb ft torque peaking at 2000rpm.

The £17,859 Shogun comes with headlamp washers and a pair of foldaway rear seats as standard. The test car has the £1250 Diamond Option Pack fitted, which adds a limited slip rear diff, electric windows and sunroof, making a total price of £19,109. On the £17,999 Citation version of the LWB Trooper, the Duty spec, which includes headlamp washers and rear limited slip diff, comes as standard. Also included is air conditioning, rear heater, central locking and electric windows.

To put the 200 Tdi Discovery on a similar equipment level, the £1050 Special Value Pack was added to the test car. This includes electric front windows, central locking, headlamp washers, electrically heated and adjustable door mirrors, twin glass sunroofs, fitted roof rack and the security pack, made up of a load space roller blind and removable storage bag. Our test car was missing both the roof rack and rear luggage cover but had the £450 uprated sound system with remote controls. Add to this £250 for the metallic paint and a further £375 for the pair of folding rear seats, and the final cost adds up to £17,875 – undercutting the Trooper by £124 and the Shogun by £1234.

DISCOVERY

nge Rover. How will it fare compared with established rivals, Mitsubishi Shogun and Isuzu Trooper?

■ OFF-ROAD TRIO

ISUZU TROOPER CITATION TD LWB

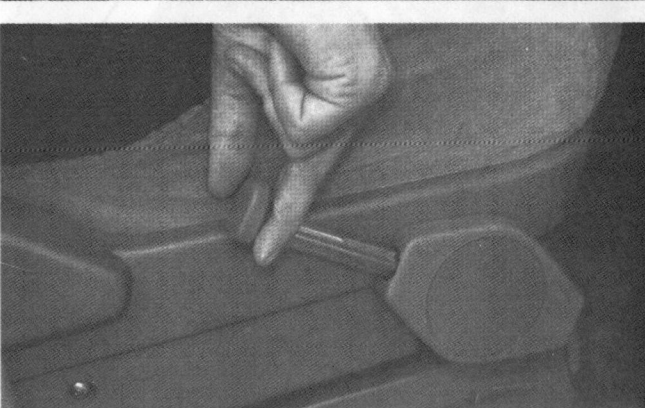

Least refined of the three, the Trooper's diesel engine is noisy and its ride is bumpy. Instrumentation is functional and can be difficult to read. Seat squabs feel rather flat, but it does have seat height adjustment. Extra rear seats cost £199

PERFORMANCE	
LAND ROVER	●●●
MITSUBISHI	●●●
ISUZU	●●

There's very little to tell these three apart when gunned hard away from a standstill. Likewise through the gears no one car stands out as being particularly slow or fast. The big difference lies in the way they deliver their respective levels of performance. The Trooper is noisy and vibrates just like a diesel of old (the new Discovery does too but to a lesser extent) while the Shogun is smooth and quiet by comparison. In fact it's so refined you'd be forgiven for thinking it a petrol engine. Clearly those balancer shafts really do make a difference.

Of the three the Isuzu is the only car that suffers from any noticeable turbo lag; during the two day photography and driving session in deepest Wales we noticed it struggling to keep up on some of the steeper ascents. Where the Shogun and Discovery would crawl happily up a hill in fourth the Trooper would be stuck between third and fourth, not being comfortable in either gear. But this you can put up with, for there are only a few occasions where it is caught out. However, the engine's lack of refinement is less tolerable. Even when fully warmed the Trooper's powerplant rattles and crackles and

WHAT'S NEW?

ISUZU TROOPER CITATION 2.8 LWB TD, £17,999: Top-of-the-range Trooper has not changed for a while. Very well specified but engine lacks intercooler and is down on power. Seven-seat capability is a £199 option.
LAND ROVER DISCOVERY 200 Tdi, £15,750: All-new mid-range car to take on Japanese at their own game. Turbodiesel engine has masses of torque – 3.5 V8 petrol available for same money. Interior very well thought out, pity only three doors. Test car fitted with £1050 Special Value Pack.
MITSUBISHI SHOGUN DIESEL TURBO 5dr, £17,859: New 2.8-litre turbodiesel engine fitted with intercooler and balancer shafts – very smooth. Test car fitted with £1250 Duty Pack to give high specification.

Isuzu Trooper Citation TD LWB, Land Rover Discovery 200 Tdi, Mitsubishi Shogun TD LWB

LAND ROVER DISCOVERY 200 Tdi

Discovery delivers best performance, but changing gear is a trial. 'Designer' interior offers excellent passenger space, but less load space. Seats give best support and there's a place for every oddment. Remote control stereo (bottom right) is an optional extra

sends a considerable amount of vibration back into the passenger compartment. Compared to the Shogun's engine the Trooper's is positively truck-like.

So much can't be said of the gearchange though this too isn't perfect, being notchy and rather difficult to rush quickly through its gate on occasion – silly as attempting to make fast gearchanges may seem in cars like these, it is in fact a fairly vital aspect during really serious off-roading, especially the change from second or third to reverse. The Shogun, in contrast, has a light, quick, and positive change with the simplest, most manageable 4wd-engage system too. The Discovery isn't so good here, its gearchange is very solid with a real feeling of heavy duty about it but some will undoubtedly find it too heavy for regular town work. The clutch is the heaviest of the threesome too, though the worst aspect is the simplicity with which you can select reverse instead of first – the gates are right next to each other, yet there's no form of prevention like a collar lift.

Though the Shogun lags behind the Discovery in outright sprinting ability, it is actually the more satisfying in a number of areas. It's smoother, quieter and suffers from less vibration which makes it the more refined of the two in most on-road situations. Off-road, however, the Discovery's low down torque advantage comes into its own and pulls it well clear of the Japanese opposition. At 1800rpm, which is nothing more than a glorified tickover, there's peak torque of 195lb ft available which gives it the sort of go-anywhere ability that only vehicles with that exclusive green and yellow Land Rover badge know.

On-road the Discovery's purpose-built engine lacks the subtlety of the Shogun's motor, but it remains comfortably within the bounds of civilisation. The Trooper's engine, on the other hand, doesn't. Get the Discovery up and rolling along a motorway and it's entirely acceptable. All you have to do is crank the stereo a little more than you would normally in your petrol burning tin box and you'd never know the difference. Around town and early in the morning particularly the crackle and vibration are more intrusive, but it's not a huge penalty to pay for the superior fuel consumption. At the end of the day we do admit that for the same money we'd specify the more powerful and vastly more refined V8 petrol engine, given the choice, despite its heavier fuel consumption.

HANDLING AND RIDE
MITSUBISHI	●●●●
LAND ROVER	●●●
ISUZU	●●

If we were judging these cars for their off-road abilities alone then the Discovery would come top, easily. Forget the

33

■ Trooper, Discovery, Shogun   OFF-ROADERS ■

MITSUBISHI SHOGUN TD LWB

Engine is smooth, ride is well controlled. Seven seats are standard in the LWB version. Storage space reasonable, but door bins too shallow. Oil pressure gauge, turn and bank indicator and voltmeter (bottom right) separated out from speedo and rev counter

fact that it doesn't have independent front suspension, it's by far the most versatile of the three when it comes to serious rough terrain. But whether this has much significance is questionable for we suspect that a great deal of Discoverys and Shoguns will never go anywhere near the mud. At the most, only once in a while just to see if it really works – as for climbing mountains, no way. Consequently it's their on-road behaviour that plays the most important part and here there is a clear winner, the Shogun. While this Japanese contender's ride is still a little bouncy and wallowy compared to ordinary road cars it is considerably better controlled than either the Discovery or Trooper. The Shogun has the best straight line stability – often a problem with such high sided vehicles – the least body roll and the most consistent ride comfort. And it's not exactly a slouch off-road anyway.

Jump into the Discovery from the Shogun and immediately you notice the heavier steering plus the thick chunky steering wheel that confirms this is no toy you're playing with. Initially the ride feels better, more absorbent because the whole car feels heavier and more solid. The first corner, however, doesn't so much shatter the illusion, but definitely changes it. The car rolls and leans far more than the Shogun but it isn't a consistent lean, just a rather vague wallow. Despite its class-leading aerodynamics the Discovery also suffers from a fair amount of buffeting in the wind or when passing lorries and buses on the motorway. On the plus side the Discovery has easily the best seats; they have more absorption built into them and offer much more side/under thigh and small of the back support. The Japanese seats feel like wooden benches by comparison.

The Trooper trails the opposition by quite a margin here. Its ride is choppy and bouncy even over good surfaces and the steering weights up badly under cornering forces, yet is too light under other circumstances. Straight line stability is marginally better than in the Discovery, but it can still be thrown off course in the same manner. Whereas the Shogun and Discovery actually have an element of refinement to their respective suspensions, the Trooper seems to have little. Overall it feels rather crude. That said we know the Trooper to be slightly more capable than the Shogun when the going gets tough off-road. But the differences are marginal and would go unnoticed to all but the hardened off-road enthusiast.

ACCOMMODATION	
MITSUBISHI	●●●●●
ISUZU	●●●●●
LANDROVER	●●●●

There is no shortage of space in any of the three, so its more a matter of which has more of it than the others. The Shogun in LWB form is a full seven-

OFF-ROAD TRIO

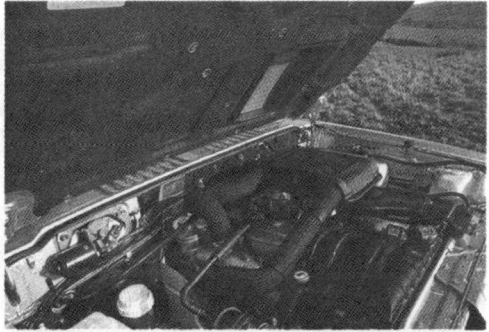
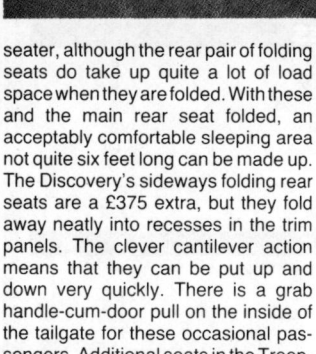

seater, although the rear pair of folding seats do take up quite a lot of load space when they are folded. With these and the main rear seat folded, an acceptably comfortable sleeping area not quite six feet long can be made up. The Discovery's sideways folding rear seats are a £375 extra, but they fold away neatly into recesses in the trim panels. The clever cantilever action means that they can be put up and down very quickly. There is a grab handle-cum-door pull on the inside of the tailgate for these occasional passengers. Additional seats in the Trooper cost a reasonable £199.

Headroom in front or rear seats is more than generous, even when sunroofs are fitted, and with the raised rear section of the Discovery's roof, back seat passengers have vast amounts of room to spread themselves in. The upright position of the front seats means that even the tallest drivers have no legroom problems. We all found that on long journeys the front seats in the Discovery and the Shogun rated as the best, with those in the Trooper feeling a little on the flat side.

We've marked the Discovery down on two counts. First is its three rather than five-door configuration. Yes, there is a very clever front seat tilt-and-forward mechanism, with a foot operated release for the rear passengers, but it is still a little awkward to clamber in and out. By contrast, the rear doors on the Trooper and Shogun make this a far more dignified process. Second is the rather restricted load space – a mere 31in from front to rear, against the Trooper's 45.5in. But the Discovery does redeem itself to some extent by being the only one of the three to have a split rear seat. We liked the Trooper's 70/30 split tailgate, which means that you can back the car closer to a big load than would be possible with the other two.

SAFETY	
LAND ROVER	●●●●●
ISUZU	●●●●
MITSUBISHI	●●●

Rule One: don't try to drive any of these cars like a sports car ... you'll eventually find out why the hard way. On the other hand, driven sensibly they are as safe, if not safer, than most other cars simply because of their size and strength. All have wonderfully high driving positions, allowing you to look over the roofs of most other cars to see what's happening ahead. Side and rear vision too is good, although the rear pillars on the Discovery are a little on the thick side. And the tailgate-mounted spare wheels do make it a little tricky to see objects obscured by them when reversing.

The Discovery scores here on two major points. It is the only car of the trio to have electrically adjustable and heated door mirrors, while the remote controls for the radio are a major safety feature. The radios on the Mitsubishi and Isuzu are placed rather low down. And who thought of putting the digital clock in the Trooper in such a position that the driver can barely see it?

There are few criticisms about the way the wipers clear front or rear screens. The switch on the Trooper is a bit tricky to operate in a hurry when a puddle empties itself over the car, and the linked wash-wipe on the test car gave just two wipes before parking. The vertical tailgates seem to attract dirt and spray like magnets, but the wiper arcs are all sufficiently large to give good visibility in foul weather. The high-pressure jet headlamp washers do a reasonable job.

None of the cars has height adjustable seat belts at the front, but the Trooper Citation does have height adjustment for the driving seat. The folding occasional seats in the Shogun have full automatic lap and diagonal belts, while the sideways-facing pair in Discovery have simple lap straps.

LIVING WITH THE CARS	
LAND ROVER	●●●●●
ISUZU	●●●●
MITSUBISHI	●●●●

As soon as you open the doors on the Discovery, you are aware of something special – even down to the owner's handbook in a Filofax-style cover to match the upholstery. Conran Design's interior gets away from the quasi-military look all too common on this type of 4wd vehicle. If there is one major criticism, it has to be the lack of a proper glove locker. Where it should be is a narrow slot, virtually filled by the bulky handbook. But this does not mean that there is a shortage of storage space. Maps? They fit into the

Though Shogun lags behind Discovery in outright performance, Mitsubishi's turbodiesel unit (top) is smoother and quieter on-road. Land Rover's torquey engine revels in the rough, where Discovery is the clear winner – but refined Shogun takes the laurels for on-tarmac work

PERFORMANCE			
	ISUZU	**LAND ROVER**	**MITSUBISHI**
Max in top (mph)	85	86	85
Max in 4th (mph)	82	81	77
0-30 (sec)	4.5	4.3	4.9
0-40	7.7	7.2	8.2
0-50	11.5	10.9	12.3
0-60	**18.0**	**16.2**	**18.7**
0-70	28.2	24.4	28.4
Acceleration through gears 3rd/4th/5th (sec)			
30-50	7.0/11.1/20.9	7.1/9.8/15.9	7.6/8.6/10.2
40-60	7.2/11.9/20.8	9.6/10.8/16.4	—/7.1/12.4
50-70	**—/15.8/22.9**	**—/13.9/21.1**	**—/16.0/17.3**
Top gear rpm at 70mph	2900	2800	3300

Isuzu Trooper Citation TD LWB, Land Rover Discovery 200 Tdi, Mitsubishi Shogun TD LWB

four slots above the sun visors, with simple wire clips to hold them in place. More maps? In the net pockets on the raised roof behind the front seats, or in the seat-back pockets. There is more room in the door bins – difficult to get into with the doors shut – or the even bigger ones either end of the rear seat and at the base of the tailgate. There is even a special holder on the back of the rear seat to take the front glass sunroof panel.

Special praise has to go to Land Rover for the facia layout. The instruments are in a binnacle ahead of the driver, flanked by press-button switches for heated rear window and rear fog lamps on the left, and those for rear wash and wipe to the right. But below these is something entirely new on a British car – remote-control buttons for the radio. You can raise or lower volume, change frequencies and then seek your favourite programme. Not quite as convenient as the Renault/Philips fingertip set-up, but not far short of it. But the switches on the column levers, borrowed from the Montego/Metro parts bin, seem a bit flimsy for this sort of vehicle. On the rough you have plenty to hang on to – for the front

passenger there are grab handles over the door and across the facia, while those passengers at the rear have handles above the side-hinged windows and built-in on either side of the front seat head restraints.

By contrast, the Isuzu Trooper's interior is rather more functional. While the rev counter and speedometer are clearly visible ahead of the steering wheel, the smaller dials for oil pressure, fuel level and engine temperature look as if they have been added as a bit of an afterthought. The flat glasses on the dials also pick up the low sun through the big side windows, making them difficult to read. Unusually for a Japanese car, there is just a single column lever for indicators, wipers and headlamp dip and flash. The small switches for the rear fog lamp and rear wash/wipe are fiddly to operate, while the big rotary one for the lights on the right of the steering column is far better. The switches for the dual-range heated front seats are set either side of the

HOW THE CARS COMPARE

CAR	ISUZU TROOPER CITATION TD LWB	LAND ROVER DISCOVERY 200Tdi	MITSUBISHI SHOGUN TD LWB
PRICE	£17,999	£15,750	£17,859
RUNNING COSTS			
What Car? test mpg	23.2	26.9	21.8
Government figures: City driving/steady speed 56/75mph	28.2/33.6/23.2	30.5/42.3/28.9	25.5/31.4/19.9
Fuel grade	diesel	diesel	diesel
Fuel tank (galls)	18.2	19.5	20.2
Range (miles)[1]	399	497	418
SAMPLE COSTS: Three years/30,000 miles			
SERVICE			
3000 miles	—	—	£27
6000 miles	£80	£20	£84
9000 miles	—	—	£27
12,000 miles	£200	£160	£153
15,000 miles	—	—	£27
18,000 miles	£85	£25	£84
21,000 miles	—	—	£27
24,000 miles	£190	£180	£245
27,000 miles	—	—	£27
30,000 miles	£85	£25	£84
SERVICE TOTAL	**£640**	**£410**	**£785**
FUEL COST	£2068	£1784	£2201
INSURANCE (Group)[2]	**£193(5)**	**£234(n/a)**	**£234(6)**
EXTENDED WARRANTY			
1 year extra cover	n/a	£247	n/a
SAMPLE DEALER PARTS PRICES			
Front brake pads	£44	£50	n/a
Windscreen	£294	£119	n/a
Exhaust system	n/a	£460	n/a
Rear lamp lens	£68	n/a	n/a
Warranty (months/miles)	12/UL	12/UL	36/UL
	24 powertrain		
Anti-rust	6	3	6
***What Car?* cost per mile (pence)[3]**	**46.5**	**n/a**	**57.2**
EQUIPMENT/OPTION COST			
Power steering	yes	yes	yes
Central locking	yes	*	yes
Sunroof	no	yes (manual)	†
Electric windows	yes	*	†
Internal mirror adjustment	no	* (elec)	no
Sound system	rad/cass	rad/cass	rad/cass
Air conditioning	yes	£1290	n/a
Limited slip diff	yes	yes	†
7 seat configuration	yes	£375	yes
* included in Land Rover's £1050 Special Value Pack † included in Mitsubishi's £1250 Diamond Option Pack			
DIMENSIONS			
Length (in)	176	178	181.1
Width (inc mirrors) (in)	75.2	78.5	77.1
Height (in)	70.9	74	75.6
Wheelbase (in)	104.3	106.1	100
Front headroom (in)	38-41	39.5	37.5
Rear headroom (in)	38.8	40.0	36.0
Front legroom (in)	33-38.5	34-42.5	33.5-39
Rear legroom (in)	25.5-32	26-35	26.5-32
Rear shoulder room (in)	55.5	59.0	57.5
Boot depth: max (in)	45.5	31.0	39.0
Boot width (in)	53.0	51.5	56.0
Boot height (in)	42.0	47.0	43.0
Boot load height (in)	28.5	26.5	28.0
Kerb weight (kg)	1815	2010	1911
Towing weight (kg)	3000	3991	3300
MECHANICAL SPECIFICATION			
Engine: cyl/cap (cc)/fuel system	4/2771/turbo-d	4/2495/turbo-d	4/2477/turbo-d
Max power (bhp/rpm)	**95/3800**	**111/4000**	**94/4200**
Max torque (lb ft/rpm)	153/2100	195/1800	173/2000
Brakes (F/R)	disc/disc	disc/drum	disc/disc
Suspension (F/R)	ind/LA	LA/LA	ind/LA
Tyres	235/75 R15	205 R16	205 R16

[1] Calculated at test mpg leaving a gallon in reserve
[2] Insurance rate based on a mean of five quotes for man aged 35 living in Oxford with 60 per cent no-claims bonus (figures supplied by Quotel Insurance Services Ltd)
[3] *What Car?* cost per mile figure calculated over three years and 30,000 miles. Includes fuel, depreciation, servicing, insurance and finance costs (figures supplied by Emmerson Hill Associates)

OFF-ROADERS

Trooper, Discovery, Shogun

awkwardly-placed digital clock way behind the gearlever. When you first drive the Trooper, the twist-to-release handbrake on the facia seems awkward, but you get used to it. The pull-up levers on the Discovery and the Shogun are far more convenient.

There is not a great deal of storage space for oddments in the Trooper, just a small, lockable glove locker and a not very large lidded box between the front seats. There are half-length and very narrow bins on the front doors, with elasticated pockets alongside.

As with the other two cars, the Shogun's driving position is best described as commanding. And like the Trooper, the instruments are separated, with rev counter and speedo in front of the driver, and with the oil pressure gauge and voltmeter in the centre of the facia, either side of the 'turn and bank' indicator. Column controls follow the Japanese Standard pattern, but the one under the main instruments can be a little awkward to get at. Storage space is fair, with a largish glove locker, and front seat map pockets. But the door bins, while long enough, are far too shallow. As with the other pair, there are map pockets on the back of each front seat.

There is not much to choose between the heating and ventilation systems of the three, all capable of pouring out huge quantities of hot and cold air. In their top forms, Discovery gets two pop-up sunroofs, the Shogun a single electric glass panel one. To make up for its lack of a sunroof, the Trooper Citation has air conditioning as standard, plus a separate fan for rear seat passengers, controlled by the driver. Discovery really does fall short of the Shogun as far as engine noise is concerned, its diesel rattle and knocking sounding rather agricultural compared with the Mitsubishi's almost petrol-like qualities.

Discovery 'designer' details (top left); V8 petrol unit is no-cost option over turbodiesel (top right). But Discovery (above) lacks extra two doors of LWB Trooper (top) and Shogun (below)

COSTS	
LAND ROVER	●●●●
ISUZU	●●●
MITSUBISHI	●●●

The prime reason for placing the Shogun last is its price. At £17,859 it's a whacking £2109 more than the brand new Discovery. Fair enough, it has an extra set of doors, seats in the back plus a more comprehensive equipment level but by specifying Land Rover's Special Value Pack you can bring Discovery up to a similar specification and still have hundreds of pounds to play with. On the plus side is a reasonable rate of depreciation – a good condition example shouldn't lose more than £1500 in two years/38,000 miles. A Trooper is likely to lose around £2500 over the same period. Plus the Shogun has that excellent three-year unlimited mileage warranty, against which the Discovery's 12 month unlimited mileage with Supercover option looks a bit sick. The Trooper's is better with 12 months unlimited mileage, 24 unlimited for the powertrain.

Servicing charges aren't yet available for Discovery, but there's no reason to expect huge bills. The Discovery and Trooper have to have an oil change every 6000 miles, the Shogun, every 3000 miles, which does reflect in their respective service totals.

The Discovery proved the most frugal at the DERV pumps returning an average of 26.9mpg, the Shogun the worst at 21.8. All have very good ranges between fill-ups thanks to enormous fuel tanks. Discovery's 19.5 gallon tank theoretically allows an amazing 524 miles before it runs dry.

VERDICT	
LAND ROVER	●●●●●
MITSUBISHI	●●●●
ISUZU	●●●

If Rover had to collaborate with Honda to develop its new 200, Land Rover can be proud of the fact that it has taken on and beaten the Japanese at the game for which they virtually wrote the rules almost on their own. Yes, the Discovery's V8 engine was originally a Buick design, while Austrian expertise was recruited for the new turbodiesel engine. But the rest is as British as roast beef and Yorkshire pud.

Knowing the exact price structure of both rivals, and aware that they are limited by the number of cars each can bring in, Land Rover has been able to aim the Discovery right at their weakest spots. It has also rewritten a lot of the rules, giving the Discovery a more attractive appearance inside and out. You cannot do too much about the aerodynamics of a brick, but by rounding the edges, the drag figure has been reduced to more acceptable levels, which is reflected in both the performance and fuel consumption.

But the Discovery does not get things all its own way. The Trooper and Shogun, in their LWB forms, have four passenger doors and huge load areas. The newcomer has just two passenger doors and places more emphasis on interior room, at the expense of load space. On the plus side, the use of an outside designer means that this Discovery is littered with map holders, storage bins and cubby holes unlike the Trooper.

With its more powerful engine, the Land Rover's performance is markedly better, but it loses out badly when it comes to noise levels and driveline refinement when compared with the Shogun. The Mitsubishi's engine feels sweet by comparison, and its gearchange is as different as chalk and cheese from the Discovery's ponderous, heavy change. The Trooper's engine suffers badly from excessive turbo lag, which means that on hilly, twisting roads, constant gearchanging is needed to keep it on boost.

Off-road, Discovery follows in the legendary wheeltracks laid down by the Range Rover 19 years ago. Its four-wheel drive system and effective but simple suspension is simply one of the best around. But on-road, the Shogun feels that bit more reassuring and steady.

If Britain can pull a car like the Land Rover Discovery out of the hat, what else is there to come?

Land Rover Discovery V8

With V8 power, the Land Rover Discovery is even better value and more than a match for Japan's best off-roaders

Price £15,750 **Top Speed** 97mph **0-60** 12.8secs **MPG** 14.0
For *Engine, space, comfort, off-road ability*
Against *Gearchange, thirst*

TEST EXTRA

Right: all-alloy 3.5-litre V8 powers Discovery to near 100mph and provides smooth and flexible delivery. Below: facia shows how it should be done. No fuss or gimmicks, just a high standard of clarity (instruments) and convenience

CONRAN DESIGN DID THE CABIN, BSB Dorland did the TV ad. The Land Rover Discovery has been well and truly launched. Only one question remains. Will Land Rover's auspicious newcomer cut a swathe through the glossy hype and do the Japanese?

For, make no mistake, its mission to end Oriental domination of the off-roader market's burgeoning middle ground could hardly be more pertinent to Land Rover's future. Ever since the Range Rover took all-wheel-driving out of the farmyard and into the fast lane, Toyota, Nissan, Mitsubishi and, latterly, Isuzu have been wallowing shamelessly in the implications.

None of the products from the Japanese giants has ever remotely approached the RR's stunning combination of off-road dexterity and King's Road chic but, to be fair, none has ever tried to. Instead, working on the 'not-everyone-can-afford-a-Rolex' theory, they've sought to provide a reasonable imitation at a reasonable price as the Range Rover's has climbed steadily skywards. Mitsubishi's well-judged Shogun has virtually cleaned up, the only serious competition coming from the younger Isuzu Trooper.

Discovery is here to stop the rot. In November's group test, the £15,750 turbo-diesel engined Tdi made short work of its oil-burning rivals from Mitsubishi and Isuzu. Now it's the identically-priced V8 petrol version's turn to step into the spotlight.

Prospective customers can't fail to notice the Discovery's most obvious handicap: just three doors. The five-door version will appear later this year though it will still be considerably cheaper than the Range Rover 3.9 Vogue, at £25,506. Kicking off with the three-door is a sound marketing ploy, anyway, allowing the Discovery's bargain basement price to hit the competition where it hurts most.

The price seems all the more remarkable when you consider that the Discovery uses what is essentially a Range Rover chassis, complete with coil springs and permanent four-wheel drive, and thus brings an altogether higher level of all-drive technology to its class. Its cabin sets new standards, too, matching Japanese rivals for space but making better use of it with clever and attractive interior design.

Obvious rivals include Mitsubishi's 3-litre V6-engined Shogun (£19,439), Isuzu's lwb Trooper Citation 2.6i 3-dr (£14,799) and Toyota's Landcruiser II 2.4 3-dr (£16,521). The cheapest Range Rover in the current line up is the £23,784 2.5 Turbo Diesel.

The Discovery V8 uses the 3.5-litre edition of Rover's venerable all-alloy pushrod V8 which, breathing through twin SU carburettors, develops 145bhp at 5000rpm and a thumping 192lb ft of torque at just 2800rpm. Heavy (4145lb) boxy and long-geared (25.1mph/1000rpm in top), the Discovery doesn't have the build of a sprinter. But with a 0-60mph time of 12.8secs and 80mph coming up in 25.1secs, it can hustle with the best of them, providing a useful step up in performance from the Tdi (17.1 and 37.1secs) and tying with the 3-litre Shogun to 60mph. Then again, the more expensive Mitsubishi has the edge on top speed, recording 102mph to the Land Rover's 97mph.

In its long-striding top gear, the Discovery covers the 50-70mph increment in a yawning 19.6secs, but the Shogun is no better, recording an identical time. Getting up to speed on motorways is best accomplished in fourth. In this gear, the big V8's fine flexibility can be more readily appreciated: 30-50mph takes 9.7secs and 50-70mph 11.7secs.

On the road, the Discovery feels no sluggard, its deeply burbling V8 delivering a relaxed and, apart from the odd induction hiccup, seamless flow of power. By off-roader standards, it's a smooth and refined unit and only starts to sound harsh above 5000rpm, beyond which it quickly runs out of puff, anyway. The engine merely murmurs at speed on the motorway — 70mph corresponds to just 2788rpm — the most prominent source of noise being wind rush.

The long gearing helps economy, too. Despite the carb-fed V8's notorious thirst, our Discovery returned 14.0mpg overall, which compares favourably with the Shogun's 15.5mpg. The projected touring figure of 19.6mpg allows a practical range of around 350 miles on an 89-litre (19.5-gallon) tankful of unleaded.

As for the gearchange, it's basically unfriendly. Extra care must be exercised when engaging first because the detent which protects reverse is pitifully weak — this, however, doesn't stop reverse being almost impossible to find when you want it. The▶

LAND ROVER DISCOVERY V8

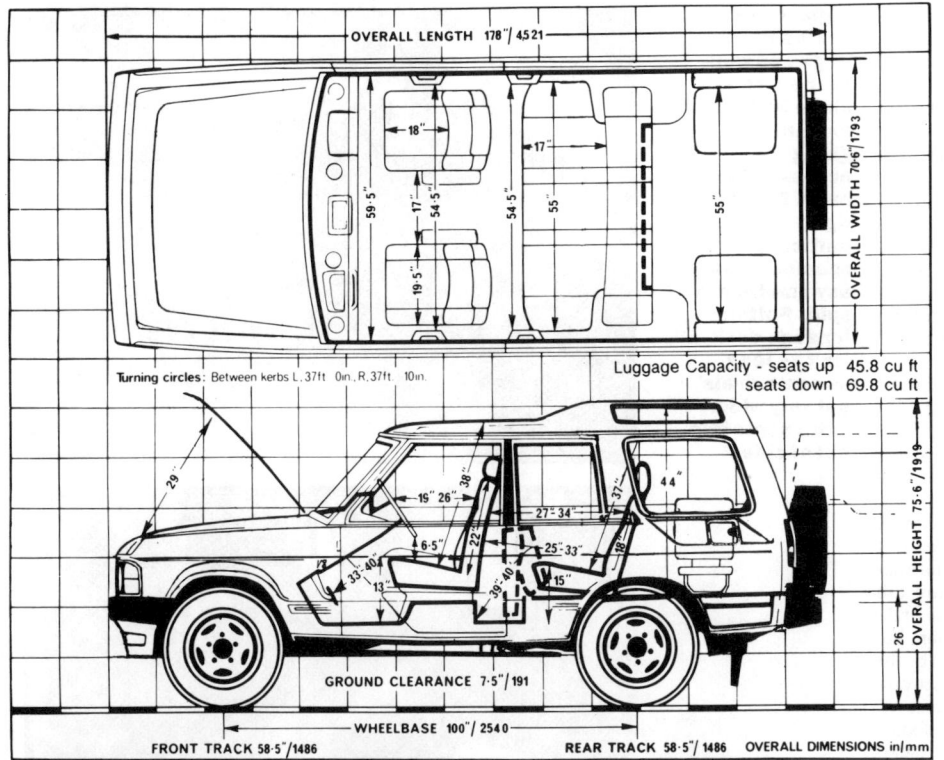

SPECIFICATION

ENGINE
Longitudinal, front, front engine/permanent four wheel drive. Capacity 3528cc, 8 cylinders 90deg V.
Bore 88.9mm, **stroke** 71.1mm.
Compression ratio 8.0 to 1.
Head/block al alloy/al alloy.
Valve gear Ohv, 2 valves per cylinder.
Ignition and fuel Electronic ignition Twin SU carburettors.
Max power 145bhp (PS-DIN) (108kW ISO) at 5000rpm. **Max torque** 192lb ft (260 Nm) at 2800rpm.

TRANSMISSION
5-speed manual.

Gear	Ratio	mph/1000rpm
Top	0.770	25.1
4th	1.000	19.3
3rd	1.397	13.9
2nd	2.132	9.1
1st	3.892	5.2

Final drive ratio 3.538 to 1. Transfer ratio: high 1.222, low 3.320, lockable centre differential.

SUSPENSION
Front, live axle, radius arms. Panhard rod, coil springs, telescopic dampers.
Rear, live axle, radius arms, upper 'A' frame, coil springs, telescopic dampers.

STEERING
Recirculating ball, power assisted, 3.8 turns lock to lock.

BRAKES
Front 11.8ins (299mm) dia ventilated discs.
Rear 11.4ins (290mm) dia discs.

WHEELS/TYRES
Pressed steel, 7ins rims. 205 R16 Goodyear Wrangler tyres.

PRODUCED BY
Land Rover Ltd
Lode Lane, Solihull,
W. Mids B92 8NW
Tel: 021-722 2424

COSTS

Prices
Total (in UK)	£15,750
Delivery, road tax, plates	£375
On the road price	£16,125
Options fitted to test car:	
Side stripes	£44
Electric pack	£525
2nd sunroof	£325
7-seat option	£375
Metallic paint	£250
Towbar	£120
Total as tested	**£17,764**

SERVICE
Major service 12,000 miles — service time 4.7 hrs. 24,000 miles service time 5.3 hrs. 36,000 miles — service time 4.7 hrs.

PARTS COST (inc VAT)
Oil filter	£10.93
Air filter	£4.72
Spark Plugs (set)	£12.88
Brake pads (2 wheels) front	£50.00
Brake pads (2 wheels) rear	£56.32
Exhaust complete	£459.75
Tyre — each (typical)	£68.77
Windscreen	£119.60
Headlamp unit	£24.50
Front wing	£111.55
Rear bumper	£80.50

EQUIPMENT

Air conditioning	£1290	
Alloy wheels	£TBA	●
Power assisted steering		●
Steering rake adjustment		●
Headlamp wash		E
Seat tilt adjustment		—
Lumbar adjustment		—
Split rear seat		●
Remote boot/hatch release		—
Internal mirror adjustment		E
Flick wipe		●
Programmed wash/wipe		●
Revcounter		●
Lockable glovebox		●
Radio/cassette player		●
Electric aerial		—
4 speakers		●
Electric windows (front)		E
Central locking		E
Tailgate wash/wipe		●
Tinted glass		●
Sunroof		●
Metallic paint	£250	●

E £525 option pack ● Standard
— Not available

WARRANTY
12 months/unlimited mileage, 3 years anti-corrosion, 12 months breakdown recovery

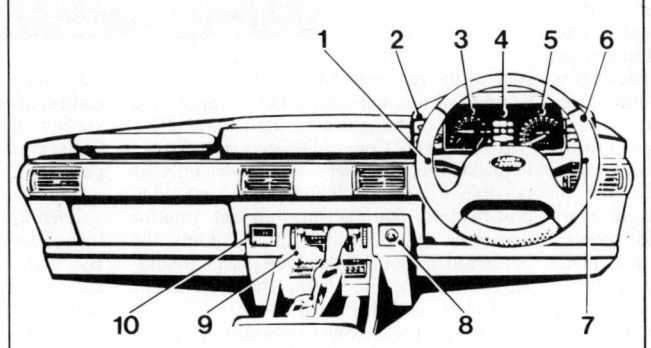

PERFORMANCE

MAXIMUM SPEEDS
Gear	mph	km/h	rpm
Top (Mean)	95	153	3789
(Best)	95	153	3801
4th (Mean)	97	156	5036
(Mean)	99	159	5109
3rd	73	122	5500
2nd	50	80	5500
1st	29	47	5500

ACCELERATION FROM REST
True mph	Time (secs)	Speedo mph
30	3.8	32
40	5.9	42
50	8.7	53
60	12.8	64
70	17.9	74
80	25.1	85
90	38.5	95

Standing ¼-mile: 18.8secs, 71mph
Standing km: 35.7secs, 88mph
30-70 Thro' gears 14.1 secs

ACCELERATION IN EACH GEAR
mph	Top	4th	3rd	2nd
10-30	18.2	11.7	7.5	4.4
20-40	16.0	10.4	6.5	4.3
30-50	16.3	9.7	6.1	5.3
40-60	18.4	10.4	7.5	—
50-70	19.6	11.7	9.7	—
60-80	21.8	13.9	11.7	—
70-90	35.4	21.4	—	—

FUEL CONSUMPTION
Overall mpg: 14.0 (20.2 litres/100km)
Touring mpg*: 19.6mpg (14.4 litres/100km)
Govt tests mpg: 13.0mpg (urban)
26.2mpg (steady 56mph)
19.6mpg (steady 75mph)
Grade of fuel: Unleaded (95RM) or (98+RM)
Tank capacity: 19.5 galls (89 litres)
Max range*: 382 miles
* Based on Government fuel economy figures: 50 per cent of urban cycle, 25 per cent each of 56/75mph consumptions.

BRAKING
Fade (from 72mph in neutral)
Pedal load (lb) for 0.5g stops

start/end		start/end	
1	20- 8	6	23-33
2	18-15	7	25-38
3	18-15	8	25-40
4	20-22	9	24-35
5	20-30	10	25-35

Response (from 30mph in neutral)

Load	g	Distance
10lb	0.22	137ft
20lb	0.45	67ft
30lb	0.65	46ft
40lb	0.68	44ft
Parking brake	0.19	158ft

WEIGHT
Kerb 4145lb/1882kg
Distribution % F/R 51/49
Test 4508lb/2049kg
Max payload 1559lb/708kg
Max towing weight 8800lb/4000kg

TEST CONDITIONS
Wind	0-3mph
Temperature	6deg C (43deg F)
Barometer	1021mbar
Surface	dry asphalt/concrete
Test distance	705 miles

Figures taken at 1543 miles by our own staff at the Lotus Group proving ground, Millbrook.
All *Autocar & Motor* test results are subject to world copyright and may not be reproduced without the Editor's written permission.

1 Indicator plus dip, **2** Heated rear screen, **3** Rev counter, **4** Warning lights, **5** Speedo, **6** Rear wash/wipe, **7** Wipers, **8** Lighter, **9** Heater, **10** Clock.

other ratios don't slip easily into place, either, for the shift action is both heavy and notchy. Persistence and deliberation pay dividends but the gearchange isn't something you can forget about. The five-speed gearbox uses the old Range Rover's gear-driven transfer box, but the whine that characterises the set-up has been well suppressed.

Soft suspension settings, huge wheel travel and large body roll angles give the Discovery's handling a recognisable Range Rover feel. The similarity extends to power-assisted steering which feels disconcertingly vague and woolly on the road yet protects the driver's hands from vicious and potentially painful kickback effects over rough ground.

On terrain that would punish lesser vehicles, the Discovery is supremely capable and remarkably easy to drive. Its full-time four-wheel drive allows the 205 R16 Goodyear Wrangler tyres to find masses of grip through mud and slime while the long-travel suspension gives unparalleled ride comfort. Generous ground clearance and terrific suspension articulation make this Land Rover a formidable rock-climber, too: with the centre diff locked, it's almost unstoppable. Engine braking isn't as great as with the diesel version, but the low range set of gear ratios provided by the dual range transfer box claw the Discovery down to a crawl in first for steep descents.

And the Discovery is amazingly surefooted on tarmac. Despite the body roll and determination to understeer, permanent 4wd once again provides plenty of grip and extremely predictable manners, especially on slippery surfaces. Lifting the throttle mid-bend merely quells the understeer, it doesn't unstick the back end. The on-road ride is less impressive *per se* than the bump-smothering act the Discovery treats passengers to in the wilderness, but it's several leagues better than a Shogun's. The all-disc braking system is powerful, progressive and largely resistant to fade though, in our tests, ultimate retardation before lock-up was a little disappointing.

The driver is well served by the Discovery. A comfortable, well-shaped seat with ample rearward travel anchors a fine, commanding driving position where the panoramic view through the screen is matched by an equally clear sight of the neat, comprehensive instruments and conveniently located switchgear. The one exception to this is the hard-to-reach five-slider heater controls carried over from the Range Rover. That said, temperature regulation is good and the centre pair of face level vents are independent of the heater.

Two sunroofs (the second an option) and a pair of 'alpine' skylights in the raised roof section give the cabin an exceptionally light and airy feel, while the rearmost seats fold down to below window level for optimum rear three-quarter visibility. Rear seat passengers are treated to plenty of legroom and, if no human cargo is present, the split rear seats fold flat to increase the volume of an already capacious luggage area. Conversely, seven can be seated by calling into play a pair of 'occasional' inward-facing seats that are otherwise stowed in the luggage area's side panels.

Cabin stowage is exceptionally well planned. As well as bins in all three doors, the facia, either side of the rear seats and the front seats, there are pockets for maps above the sun visor, net pouches in the stepped roof section at the rear and additional pouches on the back of the front seats. There's also a zipped bag on the back of the rear seats to carry the removable sunroof, while a non-slip rubber mat allows the top surface of the facia to be used as a shelf.

It's all part of Conran Design's master-plan to make the Discovery as user-friendly as possible. Conran's influence doesn't extend to post-modern minimalism, but the interior does look exceptionally unfussy. Only the predominance of powder-blue plastic mouldings — for the time being the only colour available — is slightly off-putting.

As tested, the Discovery was equipped with electric windows and mirrors and central locking but a tailgate wash-wipe and tinted glass came as standard. Our test car was fitted with the up-spec stereo featuring remote control of waveband, volume and station seek via facia-sited switches, Renault-style. This is £450 extra and, while the stereo doesn't sound particularly special, the hands-off convenience is welcome.

Even if the Discovery V8 cost £5000 more than it does, it's hard to see how it could lose. By marrying Range Rover ability with sub-Shogun pricing, Land Rover has come up with perhaps the most potent statement of all in the off-road market.

The Discovery is a real bargain, an almost perfect synthesis of flair and practicality and British-ness. A worldbeater. ■

Right: front seats are large and well-shaped. They hinge bodily forward to give good access to the back. Below right: rear is particularly spacious, seats asymmetrically split. Bottom: luggage deck isn't as long as some rivals' but very wide. Extra seats fold out of sides

New doors open for Discovery

Land Rover is hoping to sustain the Discovery's success with the addition of fuel injection and the option of two extra doors. Russell Bray reports

NINE MONTHS AFTER THE launch of Land Rover's trend-setting Discovery off-roader comes the five-door derivative, a new fuel-injected V8 petrol engine and across-the-board improvements for the 1991 model year.

Land Rover is keen to consolidate the enormous success of the three-door version, which headed the launch last November. The Discovery has, in the first six months of this year, outsold its nearest rival, the Mitsubishi Shogun, by more than two to one.

"I cannot recall any instance in recent automotive history when a market sector, heavily dominated by Japanese manufacturers, has been won back so decisively by a British company," says Land Rover commercial director Chris Woodwark.

The Discovery's easy-to-drive style and car-like qualities have proved particularly popular with estate car buyers who have never tried a four-wheel drive. "We have expanded the sector and the five-door will increase that effect," says Woodwark. Land Rover expects a 60/40 split between five and three-door versions.

Aimed at a slightly more affluent family buyer than the three-door, and so equipped with more goodies as standard, the five-door Discovery goes on sale at the Motor Show next week, when prices will be announced.

Land Rover has ditched the carburetted petrol engine in favour of a fuel-injected V8. Called the V8i and available in both bodystyles, the new power unit is derived from the old 145bhp Range Rover 3.5-litre but in a different state of tune and with a different power rating of 164bhp (153bhp if fitted with an exhaust gas catalyst).

Based on the same 100ins wheelbase as the three-door, the five-door boasts a seven-seat layout, electric windows all round, central locking, alloy wheels, headlamp powerwash, electrically-adjustable and heated door mirrors, height adjustable front seatbelt mountings and a loadspace security blind, all as standard.

The 13.2 per cent extra power of the revamped V8 transforms the Discovery's on-road performance. A 10 per cent torque improvement from 192lb ft at 2800rpm to 212lb ft at a lower 2600rpm makes overtaking quicker and safer and you can virtually drive the car like an automatic in fourth gear on a twisting main road.

LAND ROVER DISCOVERY V8i
ENGINE Longitudinal, front, four-wheel drive
Capacity 3528cc, eight-cylinders in Vee.
Bore 88.9mm, **Stroke** 71.12mm.
Compression ratio 9.35:1/8.13:1 (catalyst).
Head/block al alloy/al alloy.
Valve gear ohv, 2 valves per cylinder.
Ignition and fuel electronic ignition and injection.
Max power 164bhp (PS-DIN) (122kW) at 4750rpm/153bhp (PS-DIN) (114kW) at 4750rpm (catalyst).
Max torque 212lb ft (287Nm) at 2600rpm/ 192lb ft (261Nm) at 3000rpm (catalyst).
TRANSMISSION
Five-speed manual
Ratios Top 0.77; 4th 1.00; 3rd 1.397; 2nd 2.132; 1st 3.321. **Final drive** 3.528.
STEERING
Power assisted worm and roller, 3.4 turns lock to lock.
BRAKES
Front ventilated discs **Rear** solid discs.
DIMENSIONS
Length	178ins (4521mm)
Width	70.6ins (1793mm)
Height	75.9ins (1928mm)
Wheelbase	100ins (2540mm)
Front track	58.5ins (1486mm)
Rear track	58.5ins (1486mm)
Weight	5984lb (2720kg)

And you need to use the torque because above 4500rpm the engine sounds rather harsh and noisy.

Driving requires little effort — unless you head off-road and have to engage the stiff differential-locks lever. The power assisted steering is light but more positive than before. The clutch is heavier than a car's but it engages and disengages smoothly and the gearchange is precise.

Land Rover claims top speed is up by 5mph to 107mph, while acceleration from rest to 60mph is down from 12.7secs to 10.8secs.

Official fuel figures for the V8i are 15.2mpg round town, 24.2mpg at 56mph, and 18mpg at 75mph. The catalyst version returns 15.2mpg, 24.3mpg and 18mpg respectively.

General 1991 model-year improvements that apply to all Discoverys include flush-fitting rear seatbelt anchorages, better side window demisting and an air-conditioning option on diesel as well as petrol models.

The good news for those who don't like the 'any-colour-you-like-as-long-as-it's-pale-blue' interior is that Bahama beige is now available in addition to six new body colours. New Discovery accessories include a portable fridge/freezer, remote-control CD system and a winch.

With five doors, rear seat access is easy and there is plenty of legroom once aboard. Headroom is even better and the kick-up roof styling with Alpine windows means rear-seat passengers don't feel claustrophobic, even without a sunroof. The front seats offer good support.

The improvements display Land Rover's intention to widen the Discovery range so it can eventually replace the Range Rover with an all-new, more upmarket model. The original Discovery was good, the five-door is even better. ■

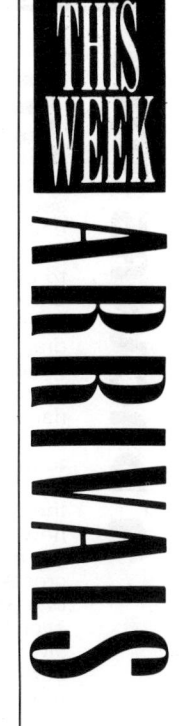

THIS WEEK ARRIVALS

Seats provide good support

There's plenty of rear legroom

GREAT DRIVES

A DROVERS TALE

Straddling the five-foot bank, with its wheels spinning uselessly, the Land Rover Discovery was now stuck fast. The climb had seemed straightforward enough at the time, but I had just been sharply reminded not to take things for granted off-road. With neither a support vehicle nor winch to drag it out of trouble, we needed to think this one out.

Steep banks flanking the narrow mountain track we wanted to join were set closer together than Britain's newest off-roader was long. Crossing the bank and ditch onto the new route would need full lock the moment the front wheels crested the rise, in theory bringing the rears diagonally over the hump. But the line that always has to be trodden between traction and momentum is a fine one, and with the very real danger of getting the Discovery wedged nose to tail between the banks if the exit speed was too high uppermost in my mind, I edged over the crest just a fraction too slowly. The Discovery's massive steel-ladder chassis grounded out before the rear wheels hit the up-slope, and almost two tons of off-roader slid to a halt.

Reversing out and trying again was simply not possible, the optional Land Rover towbar doing its usual impression of a ground anchor and refusing to allow the Discovery more than a foot or so rearwards. Rocking backwards and forwards while shuffling the front wheels from lock to lock in search of enough grip didn't do quite enough to free us – what it did do was swing the Discovery far enough to the left to lift one rear wheel up the slope. Now, with the Discovery's coil-sprung live axles right at the limit of their travel, unhindered by anti-roll bars and almost at 45 degrees to each other, we were very, very close to driving out of trouble.

Of course the Discovery was already in low ratio with the centre diff locked, but there was still a sliver of daylight under the left-hand rear wheel. Enough, with no cross-axle diff lock to prevent it, to cause both it and its diagonally-opposite front wheel to feed the generous torque of the 3.5-litre V8 into thin air. But a few deft strokes with the folding spade and we had solid ground under the rear wheel, and with the compromise tread pattern of the Goodyear Wranglers well and truly clogged we lowered the pressures to about 20psi all round for better traction. Clutch home in third gear and feeling for traction with the throttle, the Discovery inched its way over the rise and pulled up squarely on the track. Land Rover's toughest test track couldn't have provided a more serious test of the Discovery's articulation.

With lockable or slip-limited diffs at front and rear, the spadework would probably have been unnecessary. But no off-roader produced outside of Solihull provides articulation like this, and nothing with leaf springs or independent front suspension can get close on ground clearance. And if you can't keep the wheels on the

A DROVERS TALE

The first in a new series of spectacular driving stories takes Howard Lees along Welsh drovers' trails in a Discovery V8

Photography: Jim Forrest

A DROVERS TALE

The Discovery encounters its first drovers' trail

Square front end styling does little to help top speed climb much above 98mph

Challenge of Wayfarer's Trail lies ahead

ground, then all the diff locks in the world won't keep things moving – enough suspension travel and ground clearance to share available grip squarely between all four wheels is the best starting point for any serious off-roader.

Courageous step

And the Discovery is nothing if not serious. Launched by Land Rover at the end of last year, it offers Range Rover four-wheel drive technology at a very keen price and is pitched squarely into the growing market created by vehicles like the Mitsubishi Shogun and the Isuzu Trooper. Taking the Japanese on at their own game was a brave decision, and with nothing on offer in this segment before, Land Rover's Discovery would have to sell on its merits, not its reputation. To explore those merits and test the Discovery to the limit as an off-roader rather than an urban fashion accessory, we headed for the mountains of north Wales with a V8-powered version.

The intention was to follow the path of some of the oldest drover's roads over the mountains. For many hundreds of years, from after the Norman conquest to the age of the railway in the middle of the 19th century, these routes were the very lifeblood of trade between England and Wales. Hundreds of cattle, sheep, pigs and even geese in the same herd were led from the remotest Welsh farms to be sold in the markets of England by groups of Welsh drovers, most mounted on Welsh ponies and with dogs to help keep the herds together.

The drovers were fiercely independent but honest, taking livestock from farmers to sell in England, with the promise of payment on their return, and for centuries were the only means of communication between the outlying hamlets and the world at large.

That network of routes has long gone now – most either converted to metalled roads, over-planted with pine forest or submerged under reservoirs. Those that remain as green lanes can sometimes be singled out by their width – to cope with the large herds,

Close examination revealed no four-wheeler tracks any higher, just the tell-tale signs of several hard-ridden motorcycles on off-road tyres

the paths would originally have been about 14-feet wide, bounded by low banks, hedges or dry stone walls. Where they followed Roman roads, they might have been paved originally but over mountain passes, even in their heyday, the routes would have been much narrower and sometimes quite precarious.

But even if you possess the necessary very special driving skills, you can't just pick up an OS map and head off where ever you like. OS maps can

43

Wheelspin means Discovery has stuck fast, a victim of no cross-axle diff locks

show the route of many trails, but don't provide any definitive information on whether a vehicle right of way exists. The hard way to find out is to check every trail with the rights of way officer at the local council, who hold a set of definitive maps. It's easier to join a four-wheel drive club, who will be able to advise you on where you can drive, and on any routes that are particularly sensitive – it is the thoughtless use of off-road vehicles, both four- and two-wheeled, that is causing many miles of trail to be closed each year.

A wise explorer will enlist expert help, and in the Welsh mountains that help comes no more expert than Chester-based Motor Safari. Run by

> *On the limit of both grip and ground clearance, with as much momentum as I can dial up from the bridge, we just edge back on the trail*

Peter Morgan, Motor Safari run off-road training for all levels of experience at their own off-road course, but also take parties through the mountains on routes lasting anything up to a week. Peter Morgan has gone to some lengths to search out routes both suitable and legally-usable for off-roading in north Wales, though not all were originally used by the drovers – and for obvious reasons he is not all that keen to reveal their exact location.

Land droving

We intended to tackle part of one of the most important drovers' roads in the north – where the droves from Anglesey and the Lleyn peninsula would meet up with herds from Ardudwy and Harlech before heading east to Llangollen, Wrexham and Shrewsbury and perhaps as far as Kent. Gordon Downham, one of Motor Safari's senior instructors, would be our guide as we attempted the section known locally as the Wayfarer's Trail.

It's always best to venture off-road with at least two vehicles, or failing that with a decent winch bolted firmly to the front end. We had neither – what we did have was 40 years of off-road expertise distilled and blended into the Land Rover Discovery V8. It would be enough.

For the company that started the ball rolling with the original Range Rover, it has taken Land Rover a long time to react to the threat from the East. Over the past five years, Mitsubishi, Isuzu, Toyota and Nissan have made hay in Europe, with comfortable and user-friendly off-roaders at affordable prices – but by bringing coil-springing and permanent four-wheel drive to the class, Land Rover have upped the stakes.

Discovery borrows heavily from the Range Rover for its running gear. The steel-ladder chassis is almost identical, as are the live axles and coil springs at front and rear. As is Land Rover's policy, there are no anti-roll bars, but Discovery makes do without the Range Rover's Boge self-levelling strut at the rear. The drivetrain is lifted straight from the pre-fuel injected Range Rover – a 3.5-litre V8 fed through twin SU carbs and providing drive to all four wheels through a five-speed gearbox and dual-range transfer box with a lockable centre diff. On two-star or unleaded fuel, the all-alloy pushrod engine develops 145bhp at 5000rpm, with 192lb ft of torque at 2800rpm.

Like the Range Rover, most of Discovery's panels are pressed from aluminium alloy – only the stepped roof is steel – but it still scales a hefty 4200lb ready to roll. On the same 100-inch wheelbase, Discovery is a shade longer and higher than a Range Rover, but though there's only a modest increase in the space inside, it's put to much better use. Thanks to the raised roof, exterior spare wheel and a pair

Crossing the old stone bridge on Wayfarer's Trail

A DROVERS TALE

Tyre tracks scarring the climb show the Land Rover following where motorcycles have been

aerodynamics, the Discovery goes faster in fourth than top, recording a mean maximum speed of 98mph with the engine spinning just past its 5000rpm power peak. And though high gearing might help the government fuel figures, it doesn't do much in the real world – the Discovery averaged 13.1mpg in our hands, and only the generous 19.5-gallon fuel tank prevented fuel stops becoming a pain.

Through the lanes

On the twisty stuff, the Discovery has the familiar steering vagueness – a function of the hefty steering damper – and pronounced body roll of the Range Rover, though with new 205R16 Goodyear Wranglers rather than Michelins plus slightly stiffer spring rates, it feels a shade more wieldy. Once used to the alarming angles it adopts, it can be hustled along surprisingly well – there's plenty of grip, progressive understeer on the limit and, thanks to permanent four-wheel drive, little change to the Discovery's balanced stance by lifting or flooring the throttle. And on wet or slippery roads, it feels immensely secure.

We met Gordon Downham at the Bryn Morfydd hotel in Llanrhaeadr, one of Motor Safari's regular overnight halts and with an off-road training course in the fields above the Hotel. Maps consulted and route agreed, Downham, photographer Jim Forrest and I installed ourselves in the Discovery and headed for the first off-road section. This would take us over a 2000-foot

of clever folding seats in the rear, the Discovery can be a full seven-seater when needed, but at present it's a three-door only. Though the universal light blue trim doesn't look durable enough for serious use, the concept and execution of the Conran-designed interior are like a breath of fresh air. There's a tremendously roomy driving position, with well-laid out controls and enough trays and cubby holes to swallow all the usual paraphernalia and more.

Heading up the M6 then cross-country to rendezvous with Gordon Downham, the Discovery proved happy at a 90mph cruise, the engine note barely rising above a muted burble and the usual whine from Land Rover's gear-driven transfer box now well-suppressed. At that speed the tops of the doors are starting to lift away from the frames, creating a fair degree of wind noise – other Discoverys I've driven have been better in this respect. Pulling over 25mph per 1000rpm, the V8's legendary flexibility is masked in top gear, and A-road overtaking manoeuvres need a downchange or two for best results. The gearchange itself takes some getting used to, and has the familiar Land Rover weak detent between first and reverse, so with pedals widely spaced for welly-clad feet it's not a car to hurry up or down the 'box. But the brakes are first class, with a firm, progressive pedal that delivers plenty of stopping power.

For a big off-roader, the V8 Discovery is no slouch: 60mph comes up in 12.7 seconds from rest, with the quarter-mile mark despatched in 18.7 at 72mph. Limited by the high gearing and rather dubious

Following the river bed along the Maid's Trail

Resemblance to Range Rover is easy to spot

45

range of mountains and provide an indication of how the Discovery would cope with the rest of the route. At first little more than a farm track, as the climb steepened we were soon running between the low banks that characterised a droving road. With only shallow ruts and width enough to keep out of them, the Discovery was not having to flex its off-road muscles, and even at road tyre pressures the Goodyears were showing no sign of losing traction as we experimented by unlocking the centre diff.

Winding upwards between two peaks the track had narrowed, and was crossed by another linking the highest points either side of us. It was a brief diversion here that did explore the Discovery's limits. With the broken surface at the lower reaches of the climb limiting the speed at which the climb could be attempted, even third gear and the seamless torque delivery of the V8 weren't enough to stop the Discovery coming to rest part-way up the slope. Close examination of the very slippery surfaces revealed no four-wheeler tracks any higher, just tell-tale signs of several hard-ridden motorcycles on off-road tyres. But the mountain hadn't finished with us – crossing the banks back on to our original route was where the Discovery wedged itself on the brow.

Free of that obstacle, we followed the, by now, very narrow trail as it clung to the side of the mountain. There was no option but to follow the ruts here, leaving the Discovery

Painstaking progress through outstanding scenery is common to drivers and drovers alike

almost to itself and keeping thumbs out of the centre of the wheel to avoid any painful kickback that might fight its way past the very welcome steering damper. The low wire fence to our right seemed little protection against the sheer drop beyond; it can't have been much fun for the drovers, with a huge herd of cattle to look after, hundreds of years before. A final steep descent easily tamed by the V8's impressive engine braking in low-ratio first and we were back on more familiar going.

Lunch at an old drovers' inn seemed an appropriate celebration for emerging in one piece, so we retired to the Sun Inn at Rhewl. Since a drove from north Wales to Kent might take three weeks or more, the drovers were in need of plenty of overnight accommodation on the way. A group of three scots pines was the universal sign marking a farm or inn that could provide food, rooms for the head drovers and secure grazing for the animals, and some would double as a smithy, as many animals would need re-shoeing on the way. A large number of these old drovers' inns still survive throughout Wales – the Sun Inn dates back as far as the 14th century, and though it doesn't offer much in the way of grazing these days, the food is still pretty good.

Suitably refuelled, we forded a tributary of the River Dee with ease *en route* to the Wayfarer's Trail. On the Discovery, a high-mounted central air filter and intake make wading easier than in a Range Rover, and with a depth of around 18 inches and enough speed to generate the required small bow wave we crossed easily. The Discovery will handle much deeper stretches of water, but that needs some attention to waterproofing the electrics and the blanking-off of the grille with a floor mat or plastic sack.

East of the River Dee, we start to climb along the Wayfarer's Trail. This was used by the drovers travelling from the Harlech coast, meeting up along this section with the route from Anglesey through the old market town of Corwen. There are two parallel routes visible here, the northerly running through a forestry plantation that certainly wasn't there when the drovers used it. We took the more open route, the surface sticky but the ruts made by heavy logging machinery recently filled.

Keeping going

The Discovery is hugging a contour now, then dipping down to meet the second track at the edge of the forest. Here the ruts haven't been filled, so with the underside constantly brushing the ground, we rely on momentum and snatched bursts of traction as the wheels are bounced into contact with the bottom of the ruts. Even on 16-inch tyres, leaf springs or wishbones on a lesser vehicle would have dug immovably into the mud by now.

We cross a small brook over an ancient stone bridge that looks barely able to support our weight. But on the other side lies a stiffer test – a boulder-strewn set of rock steps up the hill, with sodden peat bog on either side. Picking a route here requires a lengthy recce and the careful positioning of both Gordon and Jim to mark the way, twisting between the

Climbing one of the Wayfarer's deepest gulleys

A DROVERS TALE

Conran-designed interior is clean, functional and airy. Fold-down rear seats give the option of either extra luggage capacity or seven-seater accommodation. Under the bonnet, fed through twin SU carburettors, lies an all-alloy, pushrod, 3.5-litre V8 which develops 145bhp at 5000rpm and 192lb ft of torque at 2800rpm

rocks and out on the treacherous ground on the right to avoid some differential-breaking boulders. On the limit of both grip and ground clearance, with as much momentum as I can dial up from the bridge, we just edge back on the trail above the largest rock. Only then does Gordon tell me Motor Safari call this 'no return hill' – apparently it has been some time since one of their vehicles managed the slope in an upward direction.

As the climb flattens out and the going gets easier, we are joined by another well-documented droving track from the left. Towards the 2000-foot summit of the pass, there is a simple stone memorial bearing a plaque dedicated to 'Wayfarer 1877-1956. A lover of Wales.' Erected by the Rough Stuff Fellowship, it conceals the identity of WM Robinson, intrepid off-road cyclist. Next to the stone is an aluminium box with a book to record the identity of any traveller who so desires.

Heading down from the plateau we cross seriously marshy ground, and railway sleepers have been laid across the track on the particularly difficult sections. Known locally as the Maid's Path, this was the route local girls used at harvest time to seek work in the next valley. Lower down, the track actually becomes a stream bed, though the water is not deep enough to cause any trouble for the Discovery, until we again join a farm track running down the river valley. The drovers' route now follows what are now metalled roads until it enters Llangollen – these days famous for the

> *Now, with the Discovery's coil-sprung live axles right at the limit of their travel, we were very, very close to driving out of trouble*

Eisteddfod folk festival but in droving times the first of the markets where some of the drover's livestock might be sold to English farmers.

We had spent a single challenging day off-roading in the Welsh mountains, through some of the most breathtaking scenery you could imagine. Motor Safari organise expeditions lasting up to a week, but you could easily spend a month doing the same and not be bored. Either way, you'd be hard pushed to find a more versatile ally than the Land Rover Discovery – quick, secure and comfortable on tarmac yet boasting truly outstanding off-road ability, it had tackled everything we asked of it. At a basic £15,750 it is, damn nearly, a Range Rover for the price of a Land Rover. ○

TEST EXTRA
Land Rover Discovery V8i 5dr

With two extra doors and fuel injection the Land Rover Discovery distances itself even further from the opposition

Price as tested £20,750 **Top Speed** *105mph* **0-60** *11.7secs* **MPG** *16.5*

For *Effortless performance, superb packaging, multi-role versatility*
Against *Fidgety steering, clumsy gearchange, thirst*

Discovery owes much to Range Rover concept. Interior design is excellent but colouring looks cheap

IT TOOK LAND ROVER THE BEST part of 15 years to give the Range Rover the pair of rear doors it so badly needed. With the Discovery, scourge of the Japanese, it's taken just 12 months.

The stylish and superbly equipped Land Rover Discovery V8i five-door is arguably the best interpretation of the Range Rover philosophy yet. The emphasis is on image and user-friendliness as much as on the ultimate off-road capability of a 'pure' Land Rover, and it is aimed to re-take an area of the market that has been gradually abandoned by the Range Rover owing to its escalating aspirations towards the luxury market and its consequent price rises.

The huge area in the middle of the market has been successfully tackled by the ever-opportunistic Japanese, with vehicles such as the Mitsubishi Shogun V6 and Isuzu Trooper.

Costing £20,750, the five-door Discovery is at the top end of the main marketplace for such machines (this sector is obviously not quite as price-sensitive as that for, say, small family hatches). With the introduction of five-door machines — V8 petrol and four-cylinder turbo diesel — Land Rover has made a number of useful improvements. The V8i, as tested (without an exhaust catalyst), comes as a very complete package. Sensibly, the rear compartment flip-down seats are now a standard fixture while a pair of tilt/lift-out sun hatches have been moved to the options list (£335 for both).

Mechanical changes to go with the launch of the fuel-injected 3.5-litre engine include ventilated front disc-brakes (solid rear discs retained) and a higher-ratio steering box, down from 3.8 to 3.4 turns between locks.

The majority of five-door Discovery sales are likely to be of the economy-minded 200 TDi turbo diesel, but for those desirous of a road performance falling not far short of the 3.9-litre Range Rover Vogue SE (which comes with automatic transmission as standard), then the injected V8, with manual transmission, can provide the urge.

Designed to run on any petrol down to 91-octane, the V8i runs cleanly and without the slightest protest from under 10mph in fifth gear through to a maximum of 105mph. Through the gears it passes 60mph in 11.7secs (12.8 for the previous twin SU carburettor three-door Discovery) and makes significant incremental improvements, particularly in the upper reaches of the performance range. It carves 2.9secs off the lazy driver's 50-70mph time in fifth gear (16.7secs) and lops 1.1secs off the 40-60mph time in fourth gear. The standing quarter-mile also comes up faster, at 17.7secs.

Overall fuel consumption is improved from 14.0mpg to a test consumption of 16.5mph. Our touring figure is down from 19.6 to 18.1mpg, the more powerful injection engine — 164bhp (153bhp with cat) from 145bhp — losing out in all but the Urban Cycle.

Motorway cruising is a delightfully restful affair, the V8 just bumbling along. But if you want spirited driving, then the Discovery V8i is only too happy to oblige, and the driver reaps a splendid aural bonus that only a hard-working V8 can provide.

Also enjoyable is the gear change, although the feeble detente protecting reverse needs watching as it's easily beaten when the driver is going for first. Once rolling, though, the shift is surprisingly light and snappy for an off-roader.

For off-road use the differentials can be locked and the transfer box engaged on the move to give four crawler gears that will see the Discovery climb up and down the slimiest and steepest slopes. For ultra-steep descents low first will allow it to creep down at 2-3mph with all four wheels transmitting formidable engine braking without transmission snatch. ▶

LAND ROVER DISCOVERY V8i 5DR

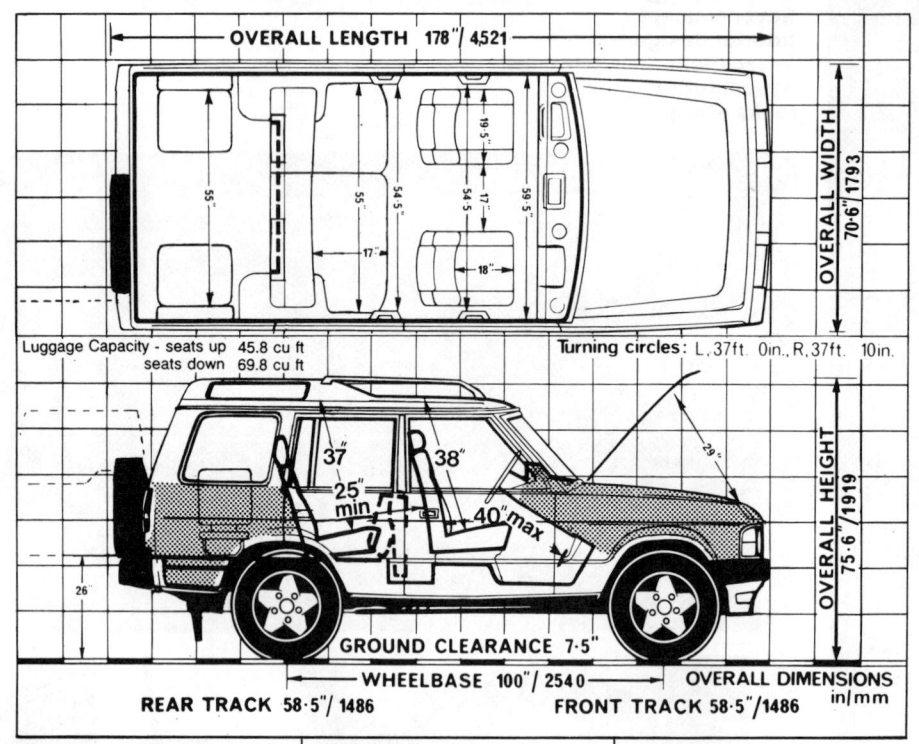

PERFORMANCE

MAXIMUM SPEEDS

Gear	mph	km/h	rpm
Top (mean)	105	169	4183
(best)	107	172	4263
4th	105	169	5500
3rd	76	122	5500
2nd	50	80	5500
1st	29	47	5500

ACCELERATION FROM REST

True mph	Time (secs)	Speedo mph
30	3.4	33
40	5.5	44
50	8.2	54
60	11.7	65
70	15.8	76
80	21.8	86
90	30.5	97

Standing ¼-mile: 17.7secs, 74mph
Standing km: 33.6secs, 93mph
30-70mph thro' gears: 12.4secs

ACCELERATION IN EACH GEAR

mph	Top	4th	3rd	2nd
10-30	15.9	10.0	6.5	3.9
20-40	14.2	9.0	6.0	3.8
30-50	14.0	8.7	6.0	4.4
40-60	14.9	9.3	6.3	—
50-70	16.7	10.3	7.3	—
60-80	20.0	11.5	—	—
70-90	27.7	—	—	—

FUEL CONSUMPTION

Overall mpg: 16.5 (17.1 litres/100km)
Touring mpg*: 17.8 (15.9 litres/100km)
Government tests: 14.8mpg (urban)
26.8mpg (steady 56mph)
19.0mpg (steady 75mph)
Fuel grade: Four star (97RM) or Unleaded (95RM)
Tank capacity: 19.5 galls (88 litres)
Max range*: 347 miles
*Based on Government fuel economy figures: 50 per cent of urban cycle, 25 per cent each of 56/75mph consumptions.

BRAKING

Fade (from 74mph in neutral)
Pedal load (lb) for 0.5g stops

	start/end		start/end
1	20-25	6	40-50
2	30-45	7	40-50
3	35-120	8	35-45
4	40-165	9	30-35
5	40-120	10	30-35

Response (from 30mph in neutral)

Load	g	Distance
10lb	0.15	201ft
20lb	0.45	67ft
30lb	0.65	46ft
40lb	0.90	33ft
50lb	1.00	30ft
Parking brake	0.40	75ft

WEIGHT

Kerb 4152lb/1885kg
Distribution %F/R 49/51
Test 4565lb/2072kg
Max payload 1839lb/835kg
Max towing weight 8800lb/4000kg

TEST CONDITIONS

Wind 10-14mph
Temperature 16deg C (61deg F)
Barometer 1011mbar
Surface dry asphalt/concrete
Test distance 953 miles

Figures taken at 2000 miles by our own staff at the Lotus Group proving ground, Millbrook.
All *Autocar & Motor* test results are subject to world copyright and may not be reproduced without the Editor's written permission.

SPECIFICATION

ENGINE
Longitudinal, front, permanent four-wheel drive.
Capacity 3528cc, 8 cylinders in 90 deg V
Bore 89.0mm, **stroke** 71mm
Compression ratio 9.4 to 1
Head/block al alloy/al alloy
Valve gear ohv, 2 valves per cylinder
Ignition and fuel Electronic ignition and fuel injection.
Max power 164bhp (PS-DIN) (122kW ISO) at 4750rpm. **Max torque** 212lb ft (287 Nm) at 2600rpm.

TRANSMISSION
5-speed manual.

Gear	Ratio	mph/1000rpm
Top	0.77	25.1
4th	1.00	19.3
3rd	1.40	13.9
2nd	2.13	9.1
1st	3.69	5.2

Final drive ratio 3.54 to 1.
Transfer ratio: 1.22:1 high, 3.32:1 low, lockable centre differential

SUSPENSION
Front, live axle, radius arms, Panhard rod, coil springs, telescopic dampers.
Rear, live axle, radius arms, upper 'A' frame, coil springs, telescopic dampers.

STEERING
Recirculating ball, power assisted, 3.8 turns lock to lock.

BRAKES
Front, 11.8ins (299mm) dia ventilated discs.
Rear, 11.8ins (299mm) dia discs.

WHEELS AND TYRES
Cast alloy 7ins rims. Pirelli Akros 205R16 tyres.

SOLD AND PRODUCED IN THE UK BY
Land Rover Ltd
Lode Lane
Solihull
West Midlands
B92 8NW
Tel: 021-722 2424

COSTS

Total (in UK) £20,470
Delivery, road tax, plates £585.35
On the road price £21,055.35
Options fitted to test car:
Metallic paint £280.00
Total as tested £21,335.35

SERVICE
Major service 24,000 miles, service time 5.3 hrs.
Intermediate service 12,000 miles, service time 4.7 hrs.

PARTS COST (Inc VAT)

Oil filter	£12.36
Air filter	£8.05
Spark plugs (set)	£14.53
Brake pads (2 wheels) front	£39.21
Brake pads (2 wheels) rear	£39.21
Exhaust complete	£315.15
Tyre — each (typical)	£90.27
Windscreen	£133.61
Headlamp unit	£26.61
Front wing	£130.87
Rear bumper	£82.94

WARRANTY
12 months/unlimited mileage, 3 years anti-corrosion, 12 months breakdown recovery.

EQUIPMENT

Anti-lock brakes	—
Alloy wheels	●
Auto gearbox	—
Power assisted steering	●
Steering rake/reach adjustment	—
Adjustable upper belt anchorage	●
Seat tilt adjustment	●
Lumbar adjustment	—
Split rear seat	●
Remote boot/hatch release	—
Internal mirror adjustment	●
Flick wipe	●
Programmed wash wipe	●
Revcounter	●
Lockable glovebox	●
Radio/cassette player	●
Electric aerial	—
4 speakers	●
Electric windows F/R	—
Central locking	●
Tailgate wash wipe	●
Driving lamps	—
Tinted glass	●
Sunroof	●
Metallic paint	£280

● Standard   — Not available

1 Indicator plus dip, **2** Heated rear screen, **3** Rev counter, **4** Warning lights, **5** Speedometer, **6** Rear wash/wipe, **7** Wiper, **8** Cigar lighter, **9** heater, **10** Clock

TEST EXTRA

◀The high, commanding driving position helps with spotting overtaking opportunities, and, down the lanes, it gives a panoramic view that can usually spot other traffic from over-the-hedge rather than around-the-corner. It inspires a safe, well-mannered driving style.

Only when pressing-on does chassis composure begin to limit speed. Using such long-travel suspension (remember that a solid front axle is employed, which has on-road geometry shortcomings compared with an independent, wishbone or strut set-up) average A or B-road progress requires fidgety steering corrections to keep a truly straight path. A steering damper, fitted to reduce wheel-fed steering shocks off-road, doesn't help with on-road feel, the overall impression being that the steering lacks precision around the straight-ahead and is too easily deflected by short and sharp road shocks. Around town, and along smooth motorways, however, the power steering is without fault.

But the Discovery is about much more than pootling along tarmac roads in the company of rank and file motor cars. Taken to the roughest tracks, the steering traits noted above come into their own. Only when the Discovery is driven over the sort of terrain that would wreck an ordinary car in the space of a few dozen yards does the Land Rover driver come to appreciate the hefty steering damper. The road wheels go dutifully up-and-down, and the body stays composed, without any shakes or rattles.

The Discovery's handling also shows off-road Range Rover parentage. On-road handling is dominated by consistent understeer and considerable body roll. Power-on cornering will eventually see both inside wheels begin to lift and lose traction. Releasing some throttle will make the cornering line tighten undramatically. All this is safe and entirely predictable but potentially off-putting for passengers, who will tend to make full use of the cleverly placed grab handles.

Off-road, the Discovery's handling is flawless. There is enough steering feel to inspire confidence, and the four-wheel-drive chassis can be manipulated by a combination of steering input, throttle (or trailing throttle) and brakes, to give near inch-perfect cornering lines over the most atrocious rough and loose tracks. The brakes are usefully set to allow the driver, should he wish, to slew the Discovery into tighter corners, when a well-judged application of throttle will see a mild transfer to controllable oversteer.

The brakes, which are so effective off-road, are less comfortable on metalled roads. Curiously, the Discovery showed signs of distress during fade testing. Pedal pressure shot off the scale (a push of more than 200lbs) accompanied by an acrid smell and clouds of smoke from within the wheels. Yet it recovered, and by the tenth stop was registering pedal pressures similar to those experienced on the first stop.

Because the Discovery V8i is so effective an off-roader, it could, perhaps, be allowed some quirks and foibles as a people carrier and multi-role carriage for those who favour green wellingtons. But it excels in these less demanding roles. The 1991 specification includes the aforementioned flip-down, side-facing sixth and seventh seats, powered windows, central locking, electrically heated and adjusted door mirrors, headlamp power-wash, height adjustable upper seat belt mountings and a pull-out rear compartment security cover.

Driving position is excellent and view panoramic. Extra rear seats are standard; fuel-injected V8 gives 164bhp

Added to this is the widely acclaimed Conran Design interior, which for fit and finish is fine, although the pale-blue colour scheme looks disappointingly cheap and can hardly be considered practical in the long term. A pale beige version is now available and, in our view, greatly preferable.

In daily use the Discovery five-door is hard to beat. Accommodation is truly first class. The sculptured rear seat (60/40 split) is perfectly acceptable for long journey use, and the flip-down load area seats (complete with safety belts) are quite comfortable enough for much more than the school run.

As for equipment, it is not found wanting. Conran Design's efforts are first class (apart from the pale colours) and from 'Alpine' windows to oddments spaces that include aircraft-style overhead magazine racks, the Land Rover exudes style combined with practicality.

The Discovery, as we last summarised it, is a world beater. Now an even better vehicle, it shows that a first-class product is being developed in just the right ways.

For the moment, the Land Rover Discovery is without equal. ■

LAND ROVER DISCOVERY V8i 5DR v TOYOTA LAND CRUISER VX v
MITSUBISHI SHOGUN V6 v RANGE ROVER VOGUE SE

For years, Japan's car giants have been trying
to build a better 4x4 than Land Rover. But they keep on failing

Will the Japanese ever win?

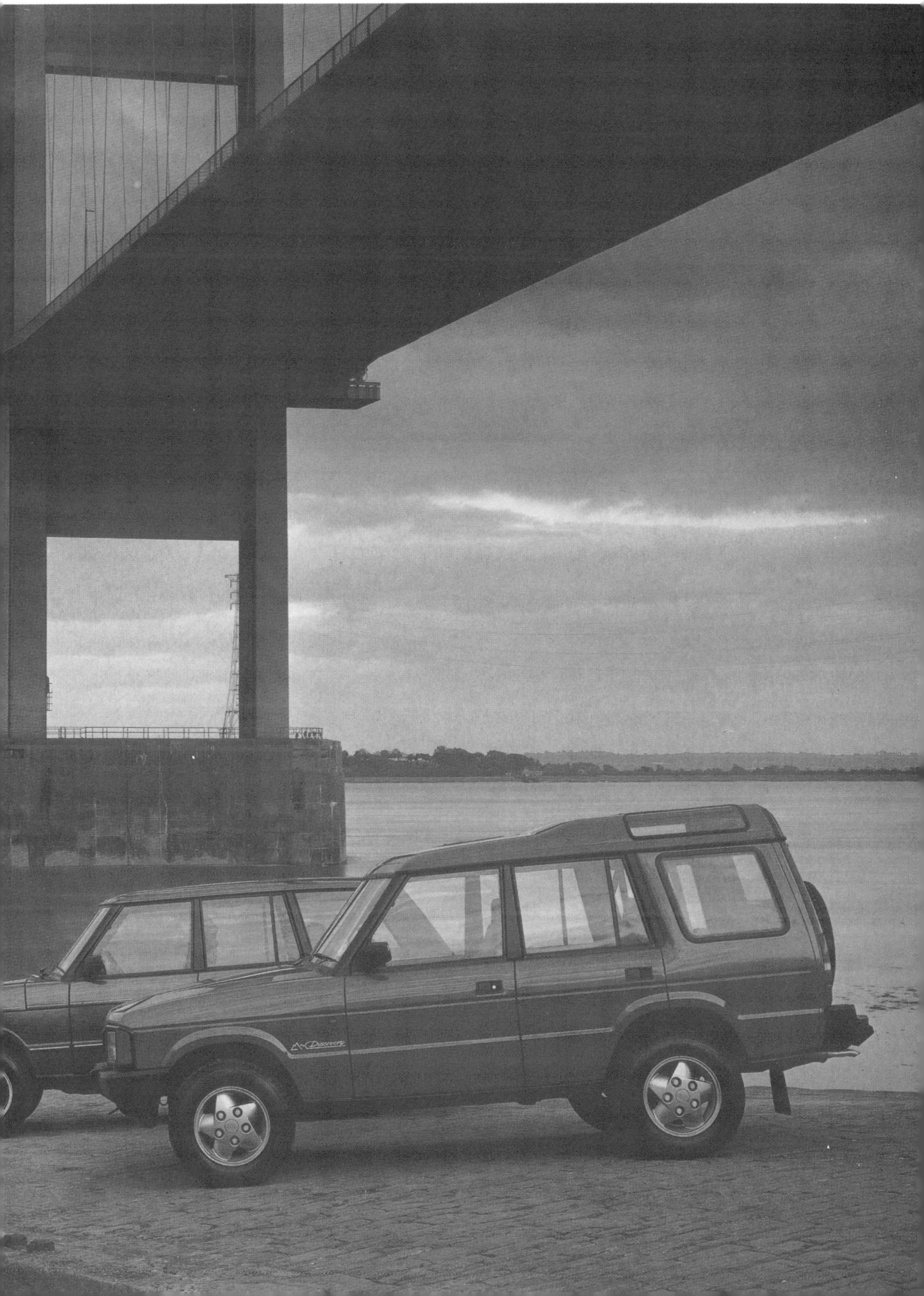
Discovery

GIANT TEST

Discovery has stiffer suspension than Range Rover, rolls less. Has more steering feel, too

Over-assisted steering and prodigious body roll make RR feel less able in bends than it is

GUCCI ACCESSORIES HAVE IT. So do 200mph supercars, Rolex timepieces and luxury off-roaders. Design excess, way beyond needs, is a powerful marketing tool, as Land Rover has proved. No manufacturer is more successful at persuading the well-heeled to spend lavishly on something they do not really need. Most Discoverys and Range Rovers are bought not because they can go where other cars can't, but because they are smart, trendy, ego-swelling vehicles that overtly reflect an affluent lifestyle. Their huge off-road ability may never be used, but its presence is an irresistible lure.

Land Rover regards the chic Discovery and opulent Range Rover as complementary, but we pitch them together here as competitors. And why not? The new five-door Discovery V8i, costing £20,470, is closer in price and power to the Range Rover than the previous carb-fed model, the appeal of which was suppressed by indifferent performance and an awkward three-door layout. That the accomplished V8i would shape up favourably to the £25,506 Range Rover Vogue, even the £28,995 limited-edition CSK, we had no doubt. But what of the £31,949 Vogue SE? Was Land Rover's high-tech flagship really worth £11,500 more?

Land Rover's opposition in this Anglo-Japanese contest had to include the £19,729 Mitsubishi Shogun V6, top model of a successful range that starts at £13,859, well below the cheapest Discovery. Although the Shogun is outsold in Britain two to one by its LR rival, Mitsubishi asserts that quota constraints distort the true picture.

At £27,741 the new Toyota Landcruiser won't worry the Discovery too much, but it falls smack in the middle of Range Rover country. If any top-end import is going to undermine the rise and rise of Land Rover – last year it made 55,000 vehicles, this year it's on target to produce a record 70,000 – it is this giant Toyonka toy from Japan.

Guided by realism, then, rather than indulgence, this confrontation was conducted not on moorland tracks but exclusively on metalled roads, the contestants' normal habitat. Take it as read that all four will clamber precipitous slopes and wallow a horsebox through sticky mud. All have enormous ground clearance, short overhangs and root-pulling ratios for bush-bashing dexterity. We are concerned here not with crawler-gear party tricks, rarely performed, but with getting around on public tarmac. Do any of these gum-booted estates really make sense on suburban gravel that's normally crunched by luxury cars?

STYLING, ENGINEERING

TRY AS IT MIGHT, THE OPPOSITION cannot match the simple elegance of the classic Range Rover. Conceived in the '60s as a fast hose-it-out workhorse, the RR has evolved over the years, without a change of suit, into the world's swankiest off-roader, a luxury five-seater that competes not with rival 4wds but, as LR sees it, with prestige saloons from Jaguar, BMW and Mercedes.

A capacity increase last year, from 3.5 litres to 3.9, pushed up power of the evergreen injected V8 engine from 165 to 185bhp. Torque jumped from 206 to 235lb ft. And it shows. Most RRs feed their considerable grunt to all four wheels through a four-speed ZF automatic gearbox, linked now to a transfer box embracing a viscous coupling which does away with the need for manual diff locks. Our test car had the alternative five-speed DIY gearbox, like that of the manual-only Discovery. Boge self-levelling and advanced anti-lock brakes – as effective off-road as they are on tarmac – are standard on the Vogue SE.

The RR's old transfer box, still with manual diff locks and whiney gears, is used in the technically inferior Discovery, which has neither self-levelling rear suspension nor ABS. Fuel injection lifts output of the 3.5-litre V8 engine by 13.2 percent, from an anaemic 145bhp to 165 (the same as the old Range Rover's), torque by more than 10 percent to 212lb ft. These outputs drop to 153bhp and 192lb ft with optional catalyst-cleaned exhaust.

Although conceived as a cheaper mid-range vehicle, the seven-seat Discovery is slightly bigger than the Range Rover. It has the same 100-inch wheelbase but is two inches longer overall, and nearly six inches taller at the back. This loftiness is needed to provide headroom for two tail-end passengers, accommodated on folding rear seats that the five-seat RR doesn't have. Instead of the RR's horizontally-split tailgate, the Discovery has a one-piece side-hinged door, carrying the spare wheel.

The seven-seater Shogun V6 longer and narrower than its Land Rover rivals, is similar in body layout to the Discovery, but it is the only car here to have part-time four-wheel drive. Normally, its 141bhp

GIANT TEST

cat-cleaned 3.0-litre V6 engine drives only the rear wheels through either a five-speed manual gearbox or a four-speed switchable automatic, as tested. All-wheel drive, in either high range or low, is engaged by operating a secondary lever, used in the other cars only to select super-low gears.

Like the Discovery and Range Rover, the Shogun is available with diesel as well as petrol power. Toyota's new eight-seater Landcruiser VX comes with an oil-burning engine only, a potent 165bhp 4.2-litre straight-six turbo that drives all four wheels through either a manual five-speed gearbox, or a four-speed auto, as tested. Not even the 3.9-litre Range Rover V8 can match the Toyota's massive torque output of 265lb ft. The broad-shouldered, heavy-lipped VX needs all the grunt it can get, mind, because it weighs nearly 2.5 tons. Hitch up a trailer, and you'll need a tachograph.

For off-road ruggedness and articulation, all four cars have well-located beam rear axles, sprung by long-travel coil springs. It's the same at the front, too, on all but the Shogun, which has a double-wishbone independent forward layout. Big tyres – none bigger than the huge 265/75 Dunlops of the Toyota – give plenty of clawing grip and ample ground clearance.

PERFORMANCE

ALL THESE LOFTY HEAVYWEIGHTS have respectable acceleration, and top speeds of around 100mph or more. None feels gutless. Fastest by far is the Range Rover, the only car here that will sprint to 60mph in under 11 seconds, and top 110mph. If that sounds close to blistering for a big off-roader, bear in mind that it's no better than average for a 1600 hatch, and poor by the standards of prestige saloons. However you look at it, weight and wind resistance seriously militate against performance and economy.

The three petrol-powered cars are unlikely to better 18mpg (all run on unleaded), and will do no more than 15mpg when pressed. The Landcruiser returned a disappointing 19mpg, its huge weight offsetting the innate frugality of its diesel engine. According to government tests, only in the urban cycle (23.3mpg for the manual, 18.5 for the auto) does the VX diesel significantly beat its petrol rivals.

There's a muscular eagerness, more GT than farmyard, about the Range Rover 3.9's crisp delivery, sharper, more virulent than that of the soft and burbly 3.5 of the Discovery. The RR's engine is smoother, too, especially around four-five, when the lesser V8 feels a bit coarse. The V8i's extra power and torque make a big difference to the Discovery's performance, lifted from lethargic to lusty by fuel injection. Economy is also marginally better. Both the long-legged LR vehicles trade low-rev muscle for effortless cruising; their V8 engines are very flexible, pulling smoothly from tickover, but they need to be extended in the intermediates to show their true mettle.

Even with lower gearing, the Shogun auto cannot match the performance of the manual V8i, never mind the Range Rover, but its sweet V6 engine sounds and feels refined. Select four-wheel drive (permanently engaged on the other cars) and acceleration deteriorates noticeably, due to extra frictional losses. There is nothing agricultural about the Mitsubishi's excellent automatic transmission, which slurs upward shifts imperceptibly, and responds keenly to down-change demands. Overdrive top can be locked out by pressing a button on the selector.

The manual shifts of the Discovery and RR are stiff, robust affairs that need firm, decisive inputs if they are to slot cleanly; keen drivers will find gearchanging rewardingly taxing, the weak or inept will not. Either way, engaging reverse is difficult, and fifth to fourth tricky on the test Discovery. The Shogun's alternative manual box, by the way, has a lighter, niftier shift. So does that of the Toyota, which goes hard for a big diesel.

Start-up clatter immediately betrays the Landcruiser's oil-burning engine, which is noisy under low-gear acceleration, and gruff when ambling. Acceleration is about as brisk as the Shogun's, top speed a bit higher, full-throttle changes occurring in D at a lowly 3600rpm, part-throttle ones much earlier. If this lazy-giant engine suffers from turbo lag, the automatic transmission – not so consistently smooth-changing as the Shogun's – effectively masks it. Once into its long-legged stride, the Landcruiser is not only effortlessly muscular, but also surprisingly quiet. There's no top-gear lockout, but it's easy to slap the unfettered lever down to third.

ROADHOLDING, HANDLING

AFTER DRIVING A NORMAL CAR, THESE tall off-roaders feel big and unwieldy. All have lightweight power steering, none true car-like agility. They are cumbersome to handle, rather than difficult, high gravity centres and long-travel suspension

Toyota has sharpest turn-in of the group and is surprisingly nimble for a car of its size

Only one of the group to have independent front suspension, the Shogun has least body roll

GIANT TEST

LAND ROVER

Although very tall, Discovery is quick cross-country. Grip is good

Conran-designed Discovery cabin best thought out, most comfortable

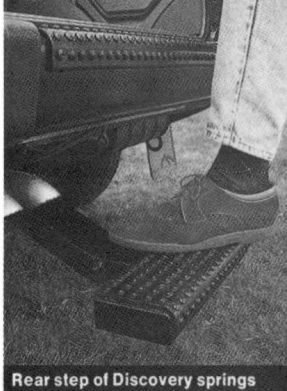

Rear step of Discovery springs out of the way when not needed

Firm, comfortable rear seats of the Discovery fold 60/40

Discovery instrument binnacle houses radio controls. Instruments clear

RANGE ROVER

Get used to the body roll, and Range Rover is a swift point-to-pointer

Clear instruments in RR, but nasty column stalks and minor switches

Panel gaps of £31,949 RR, big enough to stick fingers in

Back seat of RR shaped for two, folds down 60/40

Leather and wood cabin of RR is the classiest of the group by far

GIANT TEST

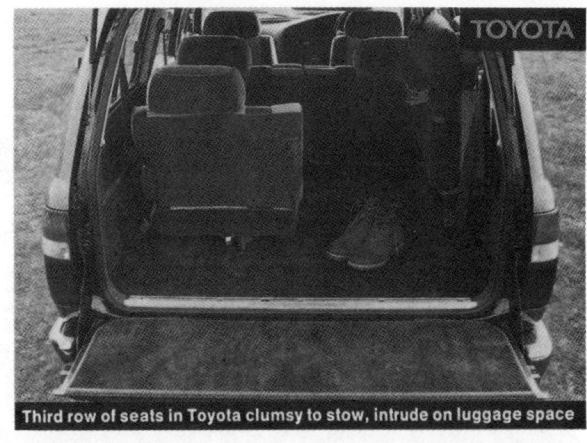

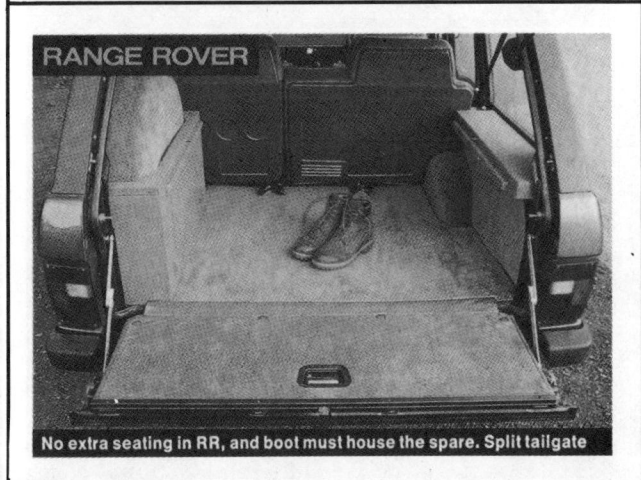

LAND ROVER — Clever seat design gives maximum boot space when chairs not needed

TOYOTA — Third row of seats in Toyota clumsy to stow, intrude on luggage space

RANGE ROVER — No extra seating in RR, and boot must house the spare. Split tailgate

MITSUBISHI — Boot of Shogun small when third row of seats erected. Narrowest cabin

inducing appreciable body lean that is not conducive to crisp handling.

Odd though they feel at first, you soon adjust to different standards, so spirited driving does not go unrewarded. Both the Land Rover products tend to wander, especially on bumpy roads, so continuous see-saw steering inputs are needed to hold line. The two Japanese cars are less prone to fidget off course; nor does their steering feel quite so vague as that of the two Brits.

RR and Discovery have essentially the same chassis, so it's not surprising they have similar handling characteristics. Of the pair, the Discovery is the tidier handler, helped by firmer springing. Its cornering behaviour is flatter, more saloon-like, and its steering inputs more precise. As with the RR, it lurches into corners when pressed, but clings on with a tenacity that inspires lots of confidence, even when leaning heavily, tortured tyres scrubbing. With lots of rubber and permanent four-wheel-drive, there's never an unsettling surfeit of power over traction.

The same can be said of the Landcruiser, which steers and corners well. Despite its weight and size, the Toyota turns into bends with more poise than either Land Rover. Its dexterity on twisty roads is really quite impressive, the grip of its huge tyres strong.

On dry roads, the Shogun corners securely, too. On wet ones, it is possible to break grip under power, causing wheelspin on tight turns (despite a limited-slip diff) and possible tail-out oversteer. All-wheel drive cannot be used to alleviate this weakness for tarmac motoring as there's no central diff to accommodate rotational differences between front and back wheels.

In normal use, the brakes of all four cars – all-disc on the LR products and Toyota, disc/drum on the Shogun – gave no cause for concern. Only the Vogue SE gets ABS.

ACCOMMODATION, COMFORT

BIGGEST VEHICLE HERE BY FAR IS THE eight-seat Toyota, more of a go-anywhere minibus than a family estate. The third row of (fold-away) seats, reached by forward-rolling the middle ones, are uncomfortable for adults. At the other extreme comes the Range Rover, designed to carry no more than five people and lots of luggage on a deck that can be extended (as it can in the other cars) by folding the back seats.

The seven-seat Discovery and Shogun have rear seats, too, those of the Mitsubishi unfurling from clumsy, strap-hanging positions. The Shogun doesn't feel as spacious inside as its rivals because it's narrower, but all four cars provide at least adequate legroom behind the front seats.

The Discovery's extra seats (optional before, now standard) face inwards, so there's more leg-stretching room than in the Shogun. They also fold away neatly into the sides, stealing little luggage space. Such clever detailing singles out the Discovery as the most well thought-out car of the quartet, as well as the most original in interior decor and design.

Up front, all four cars are very comfortable – in most respects as comfortable as a luxury car. You sit tall in sumptuous front seats, those of the Toyota being especially well bolstered for support, though a little short in under-thigh cushion. Both Japanese cars have adjustable steering wheels, but we were no less relaxed in the Discovery or Range Rover, which do not. The Vogue SE's powered seats give a wider range of adjustment than the Discovery's, but the cloth upholstery of the cheaper car supports better than the SE's slippery leather.

There's a restlessness about the way these tall heavyweights ride. They do not roar their tyres and they seldom jolt, but only on super-smooth roads do you escape the agitated head-lolling motion of rock, heave and pitch that would be unacceptable in an ordinary car.

DRIVER APPEAL

YOU GET A STRONG SENSE OF superiority from these commanding cabins. The ability to see over the roofs of cars ahead, never mind over hedges and banks, is more than a sightseeing bonus; it makes driving that much safer, particularly on twisty secondary roads.

Not even the rapid Range Rover is exciting dynamically, but all four vehicles

GIANT TEST

SPECIFICATION

	LAND ROVER	MITSUBISHI	RANGE ROVER	TOYOTA
ENGINE				
Configuration	V8	V6	V8	In-line six
Capacity (cc)	3528	2972	3947	4164
Bore (mm)	89	91	94	94
Stroke (mm)	71	76	71	100
Compression (to one)	9.4	8.9	9.4	18.6
Valve gear	Ohv	Ohv	Ohv	Sohc
Aspiration	Fuel injection	Fuel injection	Fuel injection	Fuel injection, turbocharged
Power (DIN/rpm)	164bhp/4750	140bhp/5000	185bhp/4750	165bhp/3600
Torque (DIN/rpm)	212lb ft/2600	166lb ft/3000	235lb ft/2600	266lb ft/1800
Power-to-weight ratio	81bhp per ton	79bhp per ton	96bhp per ton	70bhp per ton
TRANSMISSION				
Type	Five-speed manual	Four-speed automatic	Five-speed manual, full-time 4wd	Four-speed automatic, full-time 4wd
Ratios (mph/1000rpm)				
First	3.32 (5.8)	2.83 (6.3)	3.32 (5.9)	2.95 (7.6)
Second	2.13 (9.1)	1.49 (11.9)	2.13 (9.2)	1.53 (14.7)
Third	1.40 (13.9)	1.00 (17.8)	1.40 (14.0)	1.00 (22.4)
Fourth	1.00 (19.3)	0.73 (24.4)	1.00 (19.6)	0.77 (29.3)
Fifth	0.77 (25.1)	na	0.73 (26.8)	na
Final drive ratio (to one)	3.54	4.63	3.54	na
CHASSIS AND BODY				
Construction	Separate ladder chassis, aluminium and steel body	Separate chassis, steel body	Separate ladder chassis, aluminium and steel body	Separate ladder chassis, steel body
Front suspension	Live beam axle, radius arms, Panhard rod, coil springs	Independent, double wishbones, torsion bar, anti-roll bar	Live beam axle, radius arms, Panhard rod, coil springs	Live axle, leading arms, lateral location rod, coil springs, anti-roll bar
Rear suspension	Live beam axle, trailing arms, A-frame, coil springs	Live beam axle, lateral rod, stabiliser bars, coil springs	Live beam axle, trailing arms, A-frame, coil springs, self levelling	Four-link live axle, lateral locating rod, coil springs, anti-roll bar
Steering, type	Worm and roller, power assisted	Ball and nut	Worm and roller, power assisted	Ball and nut, power assisted
Turns, lock to lock	3.4	3.6	3.4	3.4
Turning circle (ft)	na	38.8	39.0	41.9
Wheels	7.0 x 16	7.0 x 15	7.0 x 16	7.0 x 15
Tyres	205 R16	215 R15	205 R16	265/75 R15
Brakes, type	Discs all round, vented at front	Vented discs front, drums rear	Discs all round, vented at front	Discs all round
DIMENSIONS (in)				
Wheelbase	100.0	106.1	100.0	112.2
Front track	58.5	55.1	58.5	62.0
Rear track	58.5	55.7	58.5	62.2
Overall length	178.0	181.1	175.0	188.2
Overall width	70.6	66.1	72.0	74.8
Fuel tank capacity (gal)	18.0	20.2	16.5	20.9
Kerb weight (lb)	4522	3989	4312	5302
CABIN DIMENSIONS (in)				
Front headroom	40.5	39.0	37.5	38.0
Front legroom (seat forward/back)	34.5/41.5	33.0/38.5	34.5/41.5	34.5/40.5
Rear headroom	42.5	36.5	42.5	38.5
Rear legroom (seat forward/back)	33.0/25.0	33.0/27.0	32.5/24.0	33.5/25.0
Front shoulder room	60.0	55.0	59.0	57.0
Rear shoulder room	58.0	55.0	59.0	57.0
Luggage capacity (cu ft)	45.8-69.8	na	36.2-70.8	28.7
STANDARD EQUIPMENT				
Anti-lock brakes	na	na	Yes	na
Electric windows (front and/or rear)	Yes, front and rear	Yes, front and rear	Yes, front and rear	Yes, front and rear
Cruise control	na	Yes	na	na
Sunroof (manual or electric)	Yes, two manual	Yes, electric	Yes, electric	Yes, electric
Electric door mirrors	Yes	na	Yes	Yes
Adjustable steering (reach and/or rake)	na	Yes, rake	na	Yes, reach and rake
Catalytic converter	£450.00	Yes	£449.74 option	na
Leather upholstery	na	na	Yes	na
Trip computer	na	na	na	na
Alloy wheels	Yes	Yes	Yes	Yes
MAINTENANCE				
Major service time	4.3hr	3.3hr	4.3hr	2.8hr
Oil-change intervals	12,000 miles	6000 miles	6000 miles	6000 miles
Time for removing engine/gearbox	13.5hr (incl refit)/8.2hr	3.0hr	15.6hr	na
Time for renewing clutch	6.2hr	na	5.9hr	na
Time for renewing front brake pads	0.8hr	1.0hr	0.8hr	0.8hr
Time for renewing exhaust system	1.6hr	1.5hr	1.5hr	1.6hr
Number of UK dealers	129	125	129	212
SPARES PRICES (£ EX VAT)				
Engine on exhange	2336.18 (new)	1076.52 (short)	na	na
Gearbox on exchange	1361.46 (new)	1623.33 (inc torque converter)	na	na
Clutch unit	83.40	na	180.46	171.27
Front brake disc	67.56	72.05	46.92	50.66
Set front brake pads	49.27	43.21	67.56	30.01
Damper (front)	63.62	24.54	40.96	20.04
Exhaust system	267.62	264.16 (ex cat)	364.22	185.77
Oil filter	10.75	4.28	10.56	9.95
Alternator	240.75	195.21	240.74	235.68
TOTAL COST INCLUDING CAR TAX AND VAT (£)				
Price without extras	20,470	20,259	31,949	27,401
GUARANTEE	12 months/ unlimited mileage	Three years/ unlimited mileage	12 months/ unlimited mileage	Three years/ unlimited mileage

GIANT TEST

Fuel injection gives Discovery the urge it lacked at launch – 164bhp

Turbodiesel Toyota has a plough-pulling 266lb ft of torque

Range Rover most powerful – 185bhp – but Toyota has more torque

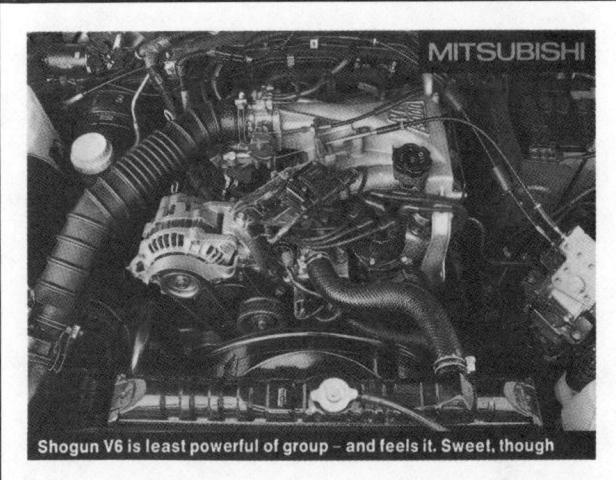

Shogun V6 is least powerful of group – and feels it. Sweet, though

are at least interesting, even quite entertaining to drive. Bulk and beefy powertrains see to that. The Discovery's appealing designer ambience contrasts sharply with the sheer ostantation of the Vogue SE. It contrasts even more with the bitza Shogun, decked out haphazardly, blocked instruments (including gimmicky tilt and altitude gauges) added almost as an afterthought. Lavish equipment does little to mask the Shogun's utilitarian roots.

For the new Landcruiser, Toyota has broken with off-roader tradition and installed a curvaceous dash that could have come from a high-performance GT. Despite its size (a major handicap on tight manoeuvres), and unfamiliar difflock controls, the Landcruiser offers drivers the most car-like environment. Somehow, it doesn't work.

CONCLUSIONS

ASSESSED OBJECTIVELY, THESE MUScular 4wds make little sense as everyday transport. If you need seven or eight seats, but not off-road prowess, a Renault Espace is a much better bet. That said, we admit to an irrational attraction to these classy juggernauts. Snob appeal apart, they are different, versatile, rugged.

For all the serious engineering endeavour embraced by the new Toyota, we cannot take the giant Landcruiser very seriously. Its marginally superior carrying capacity does not square with its great size. Huge weight also offsets the frugality of its endearing diesel engine. We liked driving the monstrous VX, but can see no role for it in suburbia, even less justification for its price.

Third place here on merit (but an easy second on value for money) is no disgrace for the poseur of the group, the Shogun V6, over-endowed with gimmicks and nudge bars, lacking a bit in cohesive design. Endearing features include a sweet V6 engine, super-smooth transmission, versatile seating (which can be turned into a bed) and build quality that out-ranks the two LR products'.

At the end of a tiring test day in Wales, everyone wanted to drive home in the Range Rover. It is the classiest and the fastest, and the nicest to sit in and conduct. It is decisively outranked, however, when money is introduced to the equation. It is simply not worth over £11,000 more than the Discovery.

Five doors and fuel injection, aided by other uplifting improvements, make all the difference to the Discovery. The V8i only narrowly wins on merit (and its interior is worthy of a design award) but it's comfortably ahead on value for money.

PERFORMANCE

ACCELERATION (sec)	0-30	0-40	0-50	0-60	0-70	0-80	30-80
Land Rover	3.4	5.5	8.0	11.5	15.3	21.6	18.2
Mitsubishi	3.8	6.0	9.6	13.2	18.9	30.5	26.7
Range Rover	3.0	5.0	7.1	10.3	13.6	18.5	15.5
Toyota	3.7	6.1	8.8	13.0	18.2	25.6	21.6
IN FOURTH GEAR (sec)	20-40	30-50	40-60	50-70	60-80		
Land Rover	9.0	8.8	9.1	9.7	11.3		
Mitsubishi	na						
Range Rover	8.3	7.9	8.0	8.2	9.1		
Toyota	na						
SPEEDS IN GEARS (mph)	First	Second	Third	Fourth	Fifth		
Land Rover	34	54	82	103.6	na		
Mitsubishi	37	69	94	na			
Range Rover	34	53	81	108	na		
Toyota	30	59	90	100.6	na		
FUEL CONSUMPTION (mpg)	Test	Urban	56mph	75mph			
Land Rover	16.1	14.8	26.8	19.0			
Mitsubishi	17.6	18.3	26.2	17.5			
Range Rover	17.5	14.4	26.9	21.1			
Toyota	19.0	18.5	27.4	18.3			

FASHION WAGONS

Land Rover has now addressed some Discovery shortcomings. So is the five-door, better-equipped, injected V8i invincible?

Discovery V8i v Shogun V6 v Landcruiser

Until recently, the off-road market had been divided into two definite camps – Range Rover and the rest. If you were in the enviable situation where money was no object, you'd take the Range Rover. But that was before Discovery.

With the launch of Discovery a year ago, Land Rover itself instantly presented the Range Rover with its biggest threat yet, despite Discovery's significantly cheaper pricing. Deliberately, Land Rover didn't initially introduce an even more direct rival, a five-door version of the Discovery, for this very reason.

Now it has, along with the option of a new, fuel-injected V8 engine, formerly carburettor only. And as if Discovery wasn't a strong enough player in the market – in turbodiesel guise, it's the 1990 *What Car?* All-terrain Vehicle of the Year – by improving and widening its appeal still further, Land Rover can't fail to make Discovery anything but an even bigger success. And to tempt the buyer further away from Shoguns, Troopers and Landcruisers, let alone Range Rovers, the factory has thrown in one or two 'extras' as standard, such as all-round electric windows, headlamp height adjustment and, significantly, the choice of a brown hue for the Conran-designed interior. Previously it's been available only in an all-pervading shade of blue.

Inevitably, these improvements mean digging deeper – £20,470 for the V8i – but it's still closer to 'the rest' on price than it is to the Range Rover. Mitsubishi's Shogun V6 5dr, for example, with its unique (for an off-roader) all-round independent suspension, smooth and punchy 3.0-litre fuel-injected V6 engine and generous seven-seat, long-wheelbase accommodation is very close on price at £20,119. But though the big Mitsubishi offers more equipment – cruise control, a huge electric sunroof plus a rather gimmicky altimeter/tilt-angle indicator – it doesn't enjoy the same image or cachet, which counts for a lot with these cars, as the Discovery.

The Toyota Landcruiser is the oddball of the trio. It uses a four-cylinder diesel engine, turbocharged in an attempt to match the petrol-powered output of the Shogun and Discovery. Nevertheless, it can't hope to rival the others on performance, majoring instead on low-rev lugging ability.

It lacks space too, being the only one here not to boast extra rear seats, simply because there isn't room for them in short-wheelbase guise. What it does have is, at £17,860, a roughly comparable price. Ideally we should have tested Toyota's brand-new VX Estate, with its enormous bodyshell and powerful straight-six turbodiesel engine, but at £27,401 it unfortunately prices itself out of the equation.

Mechanically, these three differ too. Discovery's 3.5-litre V8 produces 165bhp and a hefty 212lb ft of torque, the Shogun musters 139bhp and 166lb ft of torque from its 3.0-litre V6, while the Landcruiser's 2.5-litre turbodiesel

LAND ROVER DISCOVERY V8i 5dr £20,470

Five-door Discovery much more practical for the family. Classy, caramel-colour beats the blues, cabin gains all-round electric windows, too. A minus is stiff, notchy gearshift

MITSUBISHI SHOGUN V6 3.0 5dr £20,119

Mitsubishi sells all the Shoguns it can get, and for good reason. Suspension gives more car-like responses and a better ride than Discovery, and light major controls make it easier to drive, too

TOYOTA LANDCRUISER II £17,860

Turbocharged diesel Toyota majors on low-rev lugging ability. Rugged, chunky looks matched by bells and whistles interior. Unlike Shogun and Discovery, has no extra rear seats

OFF-ROAD TRIO

HOW THE CARS COMPARE

	LAND ROVER DISCOVERY V8i 5dr	MITSUBISHI SHOGUN V6 3.0 5dr	TOYOTA LANDCRUISER II
PRICE	£20,470	£20,119	£17,860
PERFORMANCE			
Max in top (mph)	107	102	82
in 4th	104	100	80
in 3rd/2nd/1st	76/50/28	74/46/26	62/39/21
30-70 through gears (sec)	**11.9**	**14.7**	**28.2**
0-30	3.3	3.7	5.4
0-40	5.4	5.7	8.8
0-50	8.1	9.0	13.1
0-60	**11.5**	**12.8**	**20.8**
0-70	15.2	18.4	33.6
30-50 in 3rd/4th/5th	6.0/8.9/14.0	6.3/9.3/13.7	7.1/9.8/14.8
40-60	6.2/8.6/13.9	7.1/9.8/15.3	13.1/12.9/15.4
50-70	7.1/9.1/14.6	9.3/11.1/19.6	—/19.4/27.4
60-80	—/10.0/15.9	—/12.9/22.8	—/—/—
Top gear rpm/speedo at 70mph	2800/74	3300/75	3400/74
RUNNING COSTS			
What Car? test mpg	18.7	22.3	29.1
Touring mpg[1]	17.8	19.7	23.1
Gov't mpg Urban/56/75	14.8/26.8/19.0	17.5/27.7/19.0	23.7/27.4/17.9
Fuel/capacity (galls)	UL/19.5	UL/20.2	derv/19.8
Range at touring mpg	347	397	457
Insurance Group	5	6	6
Warranty (months/miles)	12/UL	36/UL	36/60,000
Anti-rust (yrs)	no	6	6
What Car? cost per mile[2]	57.2	57.5	56.0
EQUIPMENT/OPTION COST			
Power steering	yes	yes	yes
Central locking	yes	yes	yes
Electric windows	yes	yes	front only
Adjustable mirrors	elec	no	no
Anti-lock brakes	n/a	no	no
Sound system	s.rad/cass	s.rad/cass	s.rad/cass
Alloy wheels	yes	yes	yes
Split/fold rear seats	yes	yes	yes
Cruise control	no	yes	no
Trip computer	no	no	no
Leather upholstery	no	no	no
Heated windscreen	no	no	no
Sunroof	yes	elec	yes
Adj steering column	no	yes	yes
Adj seat height/tilt	no	no	no
Adj lumbar support	no	yes	yes
Auto transmission	n/a	£540	n/a
Air conditioning	yes	no	no
Exhaust catalyser	no	yes	n/a
DIMENSIONS			
Length/wheelbase (in)	178/100	181.1/106	161/91
Width (inc mirrors) (in)	70.6	76.1	74.8
Height (in)	75.6	74	75
Headroom front/rear (in)	38/37	39.5/38	38/36
Legroom front/rear (in)	32-40/28-36	33.5-38/24-29	33-39/28-33
Rear shoulder room (in)	55	51	51.0
Boot capacity (cu ft)	45.8	46.1	12.0
Turning circle (ft)/lock turns	37.1/3.8	39.2/3.5	36.8/4.2
Kerb/towing weight (kg)	1885/4000	1900/3300	1803/1493
MECHANICAL SPECIFICATION			
Cyls/cc/fuel system	8/3528/inj	6/2972/inj	4/2446/inj turbo/d
Bore/stroke (mm)	89.0/71.0	91.1/76.0	92/92
Valvegear	ohv	sohc	sohc
Compression ratio	9.4:1	8.9:1	20.0:1
Power (bhp/rpm)	**165/4750**	**139/5000**	**88/4000**
Torque (lb ft/rpm)	212/2600	166/3000	159/2400
Brakes (F/R)	Vdisc/disc	Vdisc/drum	Vdisc/drum
Suspension front	live axle/coil	wishbones/anti-roll	live axle/coil/anti-roll
rear	live axle/coil	live axle/anti-roll	live axle/coil/anti-roll
Tyres	205R16	215R15	215R15

[1] Calculated at 'Euromix' mpg (½ urban + ¼ 56mph + ¼ 75mph)
[2] Cost-per-mile figure calculated over three years and 30,000 miles. Includes fuel, depreciation, servicing, insurance and finance costs (figures supplied by Emmerson Hill Associates)

---

trails behind with just 88bhp – but torque is a brawny 159lb ft.

Only the Discovery runs with permanent four-wheel drive, the Shogun and Landcruiser making do with a single rear-driven axle for on-road use plus a simple, manually engaged all-wheel drive system for stickier conditions.

DISCOVERY V8i 5dr

Discovery made a splash on its launch last year. Indeed, as already pointed out, we were so impressed with it as an overall package in TDi guise that we rated it best off-roader. But there were one or two vices. That turbodiesel engine, for example, is economical for such a big vehicle – around 27mpg – but lacks refinement against the likes of Mitsubishi's counter-balanced turbo-diesel motor, and creates considerable vibration when on the move.

The interior, though brilliantly designed by Conran, was available only in that rather garish shade of blue and the Land Rover lacked equipment when stacked up alongside its obvious Japanese opposition. And the final flaw was the absence of rear doors, making access to the cabin awkward.

Discovery mark two has answered all these minor criticisms in one hit, going a step further by gaining fuel injection for the V8 option. And because of its greater efficiency, the injection system actually makes the Rover-based V8 engine more economical, despite the extra power and considerably more torque.

So is the new Discovery 5dr invincible? The extra doors and improved equipment level are welcome, yet – in turbodiesel form – the underlying harshness of the engine remains. Unlike Mitsubishi's Shogun TD, the Discovery TDi is embarrassingly truck-like under the bonnet, which does little for driving pleasure.

The V8i, however, is a different story altogether. It's smooth, fast, refined and, at 18.7mpg, not outrageously uneconomical for a big, petrol-engined off-roader. Most impressive is the effortless way in which a vehicle of this size and weight performs. A 0-60mph time of 11.4sec doesn't sound terribly quick – it's about on a par with a 1.6-litre family hatchback – but on the road it feels lively and muscular.

If there are any complaints they concern driveline shunt and a notchy gearbox. The transmission jerkiness is a good deal better than it once was or early Range Rovers, yet it's annoying All that is needed to overcome this is some finesse with clutch and accelerator, but it's worth remembering that no such antics are required when driving the Shogun V6. The rather stiff gearchange is also occasionally frustrating because the gate for first is too close to reverse, and a wrong ratio can result.

SHOGUN V6 3.0 5dr

Until Discovery, the Mitsubishi Shogun was king in the *real* all-terrain market (nobody in their right mind takes £30,000-worth of Range Rover off road too often). And in top-of-the-range V6 5dr form, the Shogun is still a mighty impressive package on the road

Land Rover Discovery V8i 5dr v Mitsubishi Shogun V6 3.0 5dr v Toyota Landcruiser II

Latest Discovery makes even bigger splash: now has proper performance and a spirit all its own

Some of us preferred Shogun for its sheer ease of driving

Injected V8 more efficient, but Discovery's still thirsty

Toyota out of its depth here, but it's still appealing

Indeed, on pure drivability, it trounces the Discovery – it has an easy, precise gearshift, light steering and all-round independent suspension.

This superior suspension system provides the Shogun with more car-like responses; where the Discovery can feel heavy and cumbersome, especially when trying to conduct it tidily along a narrow B-road (which requires some skill due to the softer suspension and the vehicle's sheer inertia), the Shogun requires little more effort than is needed in a big saloon. The suspension controls the body weight far more effectively and a consequence is that the occupants are treated to a considerably more comfortable ride.

The V6 Shogun has a truly impressive equipment specification too, again far higher than that of Discovery, though it lacks the Land Rover's electrically adjustable door mirrors. Yet there is something missing in the Mitsubishi's armoury, and it strikes you the second you step inside. Somehow it lacks the Discovery's solid, quality feel, in much the same way that a highly specified Ford wants for the classy air of a poorer specified BMW. Certainly the quality of plastics used for the facia and doorcappings isn't as high as it is in the Discovery, the seats offer less support, and the switchgear is not so precise in its operation.

Yet what can't be argued about is the way the Shogun is screwed together. Even next to the well-built Discovery, the Shogun shines as being exceptional. Quite simply, it's one of the best-finished cars on the road.

Even the engine doesn't give much away to the Discovery in terms of performance and refinement – though it's in a different league to the Landcruiser's oil-burner – and the Shogun is thriftier on fuel than the British car, with a more than respectable average of 22.3mpg. The case for the Mitsubishi then – which, remember, is slightly cheaper than Discovery – is strong.

TOYOTA LANDCRUISER

The Tonka-toy, as we came to call it, is – with its chromed bumpers and over-the-top, gauge-happy interior – very much a vehicle for the American market, while for Britain it is the poseur's delight. It looks rugged and chunky and with a bit of mud on it portrays a sort of Indiana Jones of the Kings Road image.

It's not quite in the same heavy-duty off-roader league as the Discovery – it's too fragile inside for that – but it does at least look the part and, with its short wheelbase, can tackle one or two off-road obstacles that would leave the longer wheelbase Discovery and Shogun stranded.

However, on two counts it loses out to the other two here. On-road it doesn't possess anything like the same power as its rivals and thus constantly needs to be worked close to its limit if it is to keep up. Second, it doesn't have anything like the other cars' grip; those big, white-lettered tyres may look smart enough but over stickier ground they spin away much of the Landcruiser's power.

Yet for the occasional off-road driver who isn't particularly into climbing the very steepest slopes, the Landcruiser makes a lot of sense. Four-wheel drive is engaged simply by the flick of a switch on the right of the dash, and the gearshift is quick and easy. Indeed, viewed in isolation, the Landcruiser is an immensely appealing and different means of everyday transport. It's comfortable enough (though the on-road ride is definitely inferior to that of the Shogun and Discovery), it can just seat four and there's plenty of equipment – including an altimeter and tiltometer to entertain the passengers.

It's only when you put the Toyota alongside the Shogun V6 and Discovery V8i that the shortcomings fall into view. Though it isn't that much cheaper – £2610 less than the V8i Discovery and £2259 short of the Shogun V6 – it offers the buyer considerably less. Space, performance, 'image' and refinement are all in shorter supply. It does make up some lost ground with its superior fuel economy – 29.1mpg against the Shogun's 22.3mpg and Discovery's 18.7mpg – and lower asking price, but overall it is the poor relation of the three.

RATINGS AND VERDICT	L/ROVER	MITSUBISHI	TOYOTA
PERFORMANCE & ECONOMY	●●●●○	●●●●○	●●●○○
HANDLING & RIDE	●●●●○	●●●●●	●●●○○
BEHIND THE WHEEL	●●●●●	●●●●●	●●●○○
ACCOMMODATION	●●●●●	●●●●●	●●●○○
QUALITY & EQUIPMENT	●●●●●	●●●●●	●●●●○
SERVICING & COSTS	●●●○○	●●●●○	●●●●○
VERDICT	●●●●○	●●●●○	●●●○○

Despite its improvements, the Discovery is run very, very close by the Shogun. Indeed, for one tester the Mitsubishi took top spot. Why? Because it's considerably easier to drive on the road (though it does lack the Discovery's ultimate off-road ability), requiring less effort to change gear, to corner properly and to drive in a straight line.

Yet the Shogun does lack the Discovery's rugged, hewn-from-the-solid appeal, and the Land Rover has an altogether more classy atmosphere inside. That charismatic V8 growl does much to win you over too, endowing the latest Discovery with proper performance and a spirit all of its own. The fuel economy isn't marvellous, but it's considerably better than it was. Overall, it just tips the Shogun into second place – but by a smaller margin than we imagined.

As for the Toyota, it's pleasant enough – fun, even – in its own right, but up against the two best in class it really can't compete. Nevertheless, a Toyota strongpoint is the good fuel economy.

Overall though, it's easy to see why the Discovery is an ever-more-common sight on British roads – and why Mitsubishi can sell every Shogun it can lay its hands on.

RUNNING REPORT

HARD AT WORK

Martin Vincent has been making full use of a rejuvenated Discovery's go-anywhere ability

THESE DAYS, OUR LAND Rover Discovery TDi is starting to work a little harder for its living. Off-roaders can be such slow, noisy and thirsty beasts on the road that it makes sense to exploit at least some of its versatility over different types of terrain.

That's how I came to find myself peering over the Discovery's bonnet with nothing but wide blue yonder beyond. Much against my better judgement, I was teetering on the edge of some precipitous quarry workings, urged onwards by experienced off-roader Peter Morgan. "First, make sure you're square on, then just release the clutch, leave it in low-range first, and keep your feet clear of the pedals and with the steering in the straight-ahead position. Just let the machine do the work. It's easy."

And it was, too. The front dipped down over the edge, just scraping the chassis in the centre, to reveal a steep and slippery slope of perhaps one-and-a-half car lengths, and a short flat area beyond, followed by an even steeper slope leading to the treetops lurking just over the edge. The powerful braking effect of the turbo diesel engine, together with the ultra-low gearing, kept the speed of descent down to a level that permitted a turn at the lower end of the slope.

"It's so easy to panic and to stomp on the brakes and the clutch just as you gather momentum, but that's the worst thing you can do," observed my mentor. Peter Morgan knows about these things. He runs the Motor Safari off-road course, and he was attempting to give me a crash course in off-road driving, but without the crashes.

The remainder of the day was crammed full of intensive training on Peter's compact but varied course. Ascending steep and slippery trackways, I learnt to saw the steering wheel from side to side to gain traction from the sides of the ruts to maintain momentum; I learned how to stall and reverse back down slopes that the vehicle is unable to conquer; I practised throttle control to prevent excessive wheelspin; I learnt how far an off-roader can be leaned in safety while traversing a hillside, how best to plan the approach to obstacles and how to make progress with the minimum damage to the environment.

All the time, the Discovery proved everything that we already knew about its off-road competence. It's a supremely comfortable and capable machine, willing and able to cross the toughest of terrain.

It wasn't long before I was looking for the opportunity to try out these new-found off-road skills, albeit in a somewhat less harsh environment. Just a few miles from home is the Oxfordshire Ridgeway, 30 miles or more of ancient droving road. In some places it's almost as broad as a motorway, but in others the hawthorns close in and the ruts deepen to make it a tight squeeze for a Discovery.

It's a trail I know well, and one that I go back to again and again for its solitude and the wide open spaces. In the past, I've travelled the Ridgeway on two wheels rather than four, first with my trusty but ancient 1949 AJS but later on a far more potent BSA Victor, an ex-motocross machine with silencer and number plate added.

These days, when I hanker for a spot of off-roading, my two young children want to come, too. And that's where the Discovery comes in handy.

After piling the boot up with a picnic, various pairs of wellies, a shovel (in case we get stuck) and a kite or two, we joined the Ridgeway at Streatley, pausing after a few miles to watch the racehorses training on the gallops, then bounced along for another 10 miles until the White Horse hill at Uffington appeared alongside. Then out came the picnic and the kites, followed by a walk around the bronze-age hill fort with its spectacular views.

If it's raining, no matter — we have the picnic in the Discovery. Its vast areas of flat, rubber-matted oddment shelf are ideally suited to the job, and the wipe-clean plastic is easy enough to clean after the inevitable spills of coffee and smears of mud. At weekends, the Discovery is always in demand.

Since its last service, carried out at 24,000 miles, the Discovery is at last back on form. All the drivetrain backlash and the noisy clonks when coming on and off the throttle have at last been banished. The culprit? A worn ball joint coupling for the rear axle. This had previously displayed all the symptoms of a seriously worn drivetrain. Full marks to Guy Salmon of West Ewell for locating the fault, and also for effecting a permanent cure for the errant diesel return pipe unions. Lock-wiring was the solution.

A small modification has banished the idler gear chatter, although this was never a serious complaint, and the worn fan belt was replaced. At the same time, the front brake pads and the thermostat were replaced and the rear door was readjusted. After its last service, the Discovery feels like it's just come back from a stint at a health farm. And it is at last possible to drive smoothly again, thanks to the replacement ball joint. But guess what? The rear door has dropped yet again. There's no denying the Discovery has its little faults and foibles, but its likeable character and inspired design are ample compensation.

LAND ROVER DISCOVERY TDi

Acquired	March 1990
Total mileage	28,700
Original price	£18,168
Overall mpg	26.7
Best/worst mpg	31.9/20.6

Faults and failures
On delivery: None.
600-1400 miles: Tailgate latch fell off.
1400-10,000: Broken rear window latch, failed ignition switch.
10,000-15,000: Fuel injector pipes came loose, one detached, sunroof catch broke, gear lever knob worked loose.
15,000-24,000: Rear door dropped on its hinges, electric mirror adjustment failed, fuel injector pipe came loose again.
24,000-28,700: Rear door dropped on its hinges again.

OFF-ROADERS

BOG STANDARDS

All three cars are capable off-roaders; the Discovery has the best gear ratios and articulation, while the Landcruiser has the strongest engine and the G-Wagen is superbly built

We've been full of praise for Land Rover's Discovery in the past, but how does the turbo diesel model rate against the Toyota Landcruiser and Mercedes G-Wagen? Find out in our three-way shootout. And we provide tips on buying and enjoying a 4wd

LAND ROVER POLE-AXED THE OFF-road opposition when it unleashed the Discovery. With its Range Rover-derived chassis but more modern design, and priced to hit the Japanese imports where it hurt most, the hugely successful Discovery won plenty of customers as well as *Autocar and Motor's* 'Top Off-roader' title two years running. In the eyes of many it fully justified its Range-Rover-on-the-cheap image.

But on the other hand, few doubted that the Japanese would waste much time in fighting back. The 1991-spec Toyota Landcruiser — now with permanent four-wheel drive and electronically operated diff locks — totes the sort of technological hardware to cast an ominous shadow over the Discovery's talents. A similarly sophisticated drivetrain graces the re-engineered Mercedes G-Wagen. Both are after the Discovery's blood.

The Discovery seeking to beat them off in this confrontation is the turbo diesel Tdi in five-door form. It's also available with three doors (on the same 100ins wheelbase) and with petrol V8 power. Listed at £19,785 (£21,390 as tested) the Discovery Tdi five-door wields a significant price advantage over the better equipped Landcruiser and G-Wagen. Both of these imports come in short-wheelbase three-door or long-wheelbase five-door form, but although the G-Wagen can be ordered with either a 170bhp petrol six or a normally aspirated diesel, Toyota's long-wheelbase Landcruiser VX is available (in the UK) only with a 4.2-litre six-cylinder turbo diesel. The Japanese seven-seater is massive in every sense — not least in its torque output (265lb ft) and price (£27,401). But the Mercedes long-wheelbase diesel, the 300GD, is more expensive still, at £31,890. Add the test car's extra equipment and you're looking at £34,436, which makes the Land Rover appear cheap.

We had a formidable challenge in mind for these tough and diesel-powered 4×4s. After performance testing at the Millbrook test facility, we headed for the Oxfordshire Ridgeway for 30 miles of mud, slime and ruts. Then it was on to the motorway and into Wales, where again the 4×4s got down to it in a completely different type of rocky, precipitous terrain. Finally, we headed back towards London using a mix of A and B roads.

AT THE TEST TRACK

All three of these 4×4s are diesels, but that's where the similarity in engines ends. The Discovery uses a Land Rover-designed 2.5-litre turbo diesel four (not the VM turbo diesel of the Range Rover), but both the others have straight sixes new to these applications. Mercedes has switched to the 3-litre normally aspirated engine lifted from the mid-series saloon range, while the Toyota's turbocharged overhead cam six is all new and, at 4.2 litres, bigger than the one it supplants.

Indeed, it's hard to argue with a walloping 265lb ft of torque (developed at just 1800rpm) and 165bhp at 3600rpm. The Discovery tries hardest, armed as it is with 195lb ft of torque at 1800rpm and 111bhp at 4000rpm but, alas, the G-Wagen doesn't have a hope. Its 141lb ft at a fairly high 2900rpm and 113bhp at 4600rpm aren't enough to propel a two-and-a-quarter-ton 4×4 with the wind-cheating properties of a barn door at a respectable pace. Not that the Landcruiser and Discovery are much lighter, and neither seems to have seen the inside of a wind tunnel. ▶

Not surprisingly, the Landcruiser is the only one capable of cracking 100mph. In fact, at 103mph its top speed is 16mph clear of the 87mph recorded by the Discovery and far ahead of the G-Wagen's 84mph.

If car-like performance and diesel power are requirements, only the big-engined Toyota will do. Indeed, the Landcruiser's 12.6secs 0-60mph time is in the large 2-litre estate car league, and it's not that far off the figure achieved for the V8 petrol-engined Discovery. If the diesel-engined Discovery is tediously slow, recording 19.2secs to 60mph, then the Mercedes is absurdly so. High revving its engine may be, but to little effect: the G-Wagen can't even break 25secs to 60. Between 30 and 70mph through the gears the order is the same: Toyota 13.9secs, Land Rover 23.8secs, Mercedes 33.9secs.

After this, we wondered if it might be asking too much of the G-Wagen to provide top gear acceleration times, but we tried it anyway, and recorded 50-70mph in a soporific 42.5secs. More than 80mph was beyond it on Millbrook's mile straight. A hopelessly optimistic speedo couldn't influence matters here, even if it does give a false impression elsewhere. The Toyota dispatched the same 50-70mph increment in a bullish 11.5secs, the Land Rover in a middling (disappointing by absolute standards) 24.8secs.

ON THE MOTORWAY

There's an argument for making the motorway the Toyota's preferred domain. Its effortless torque is relaxing, while the subdued quietness and lack of diesel harshness renders lengthy journeys painless.

But although the G-Wagen equals the Toyota for engine smoothness and overall refinement, wind roar is ever-present, and its bluff facade and lack of power ensure a constant struggle to keep up with the traffic flow in a headwind or going uphill.

While the Discovery makes lighter work of motorway travel, it too struggles on inclines, and it is by far the least successful at isolating vibration, noise and harshness from the engine and drivetrain. On long journeys this constant vibration and the deep, rumbly diesel clatter are wearing and unpleasant. Also, the steering constantly demands small inputs to counter its fidgety waywardness. All three cars react adversely to crosswinds.

The Discovery claws back some self-respect by dint of its better control over undulating surfaces. In fact all three ride with reasonable suppleness on the undemanding environment of the motorway, but the Discovery irons out the imperfections slightly better than the Toyota. The G-Wagen controls big undulations with ease but finds it more difficult to absorb the sharper bumps and ridges, which create a shudder through the structure. Its soggy, ineffectual brakes count against it too.

AROUND TOWN

It's when approaching the tight confines of urban streets that the sheer bulk of a big off-roader becomes abundantly clear. Manoeuvring in tight spaces is an effort, but the elevated driving position, good all-round visibility and power steering common to all of these offer some compensation.

In town, the Toyota's ultra-light controls make it easier to drive than the Discovery, which suffers from a stiff gearchange and heavy clutch. And both the G-Wagen and Landcruiser are quieter and freer of vibration than the comparatively harsh and noisy Discovery. All are cumbersome to park.

At town speeds, the G-Wagen's slothful progress is less obvious than when out on the open road, but the heavily damped throttle — which makes the revs soar between gearchanges unless the pedal is released early — makes driving smoothly harder than it should be. Don't expect to be able to beat anything quicker than a 2CV away from the lights with the Benz.

The Discovery copes best with absorbing the worst neglect of urban streets, but genuine large-car ride standards simply aren't on the agenda. All suffer from an underlying propensity to judder over sharp-edged imperfections and ripples.

ACROSS COUNTRY

Despite its jumbo proportions the Landcruiser is the easiest of the three with which to attack give-and-take roads in search of enjoyment. Not only does it have the reserves of power available for safe overtaking, but firm and positive brakes, reasonably precise steering, acceptable grip and a safe cornering balance.

Push things too hard, however, and shortcomings are apparent. Steering becomes light and soggy and previously progressive roll degenerates into unseemly lurching. Clearly brisk tarmac bend-swinging was never intended to be a part of the Landcruiser's otherwise well rounded repertoire.

Paradoxically, it's the oh-so-slow G-Wagen that steers with the most accuracy and feedback, but everything is relative. By almost any other standards the steering of any of these would be classed as sloppy.

That's certainly the initial impression with the soft-sprung Discovery. It rolls to a massive degree. But once committed to a line, the steering wakes up and feels both communicative and surprisingly accurate. The Discovery is the most secure, with the best grip and most positive brakes of the lot, despite being equipped with the narrowest tyres.

What is clear is that the 'G' in G-Wagen doesn't stand for 'grip'. It skates towards the outside of a bend far too readily and the poor adhesion is largely responsible for its ineffectual braking, too. It is the only one equipped with anti-lock brakes, but its maximum retardation of around 0.5g before tyre lock-up (admittedly on wet tarmac) can only be described as pitiful.

As already established, limp-wristed performance is another aspect of the G-Wagen that fails to inspire. Forget about overtaking even the slowest of trucks unless there's a straight disappearing over the far horizon. The G-Wagen driver is destined for a life of staring at tailgates. The Discovery driver has better engine response at his disposal, though passing slower traffic still requires planning.

None of the trio possesses a gearchange quality that provides much reward. The over-long shift of both the Mercedes and Toyota, and the treacly action of the Discovery's lever, put the lid on any potential enthusiasm for changing gear.

OFF THE BEATEN TRACK

The first off-road arena for our trio was the mud and ruts of the Ridgeway. One of the most famous ancient rights of way in the country, ▶

HOW THEY COMPARE

	DISCOVERY Tdi	LANDCRUISER	G-WAGEN 300GD
ENGINE			
Type	turbodiesel, 4cyl	turbodiesel, 6cyl	diesel, 6cyl
Capacity, cc	2495	4164	2996
Bore/stroke, mm	90/97	94/100	87/84
Max power, bhp	111	165	113
Max torque, lb ft	195	265	141
Power/weight ratio, bhp/ton	56	74	50
GEARBOX			
Mph/1000rpm in top	25.1	26.6	20.5
Tyres	205×16	265/75×15	255/75×15
DIMENSIONS/WEIGHTS			
Length, ins	178.0	188.2	173.0
Width	70.6	74.8	66.9
Height	75.9	74.4	77.8
Wheelbase	100.0	112.2	112.2
Max ground clearance	8.1	8.7	8.2
Kerb weight, lbs	4427	4971	5051
Fuel capacity, galls	18.0	21.0	20.9
Overall test mpg	23.3	19.8	17.1
PERFORMANCE			
Max speed, mph	87	103	84
0-60mph	19.2	12.6	25.4
30-70mph	23.8	13.9	33.9
Standing ¼ mile	21.3	19.2	23.4
Acc in third			
20-40mph	7.2	5.9	9.5
40-60mph	11.4	6.1	—
Acc in fourth			
30-50mph	10.1	9.1	14.7
50-70mph	16.5	9.3	24.2
Acc in top			
20-50mph	17.1	11.8	23.4
50-70mph	24.8	11.5	42.5
Price	£19,785	£27,845	£31,890

MERCEDES G-WAGEN 300GD LWB

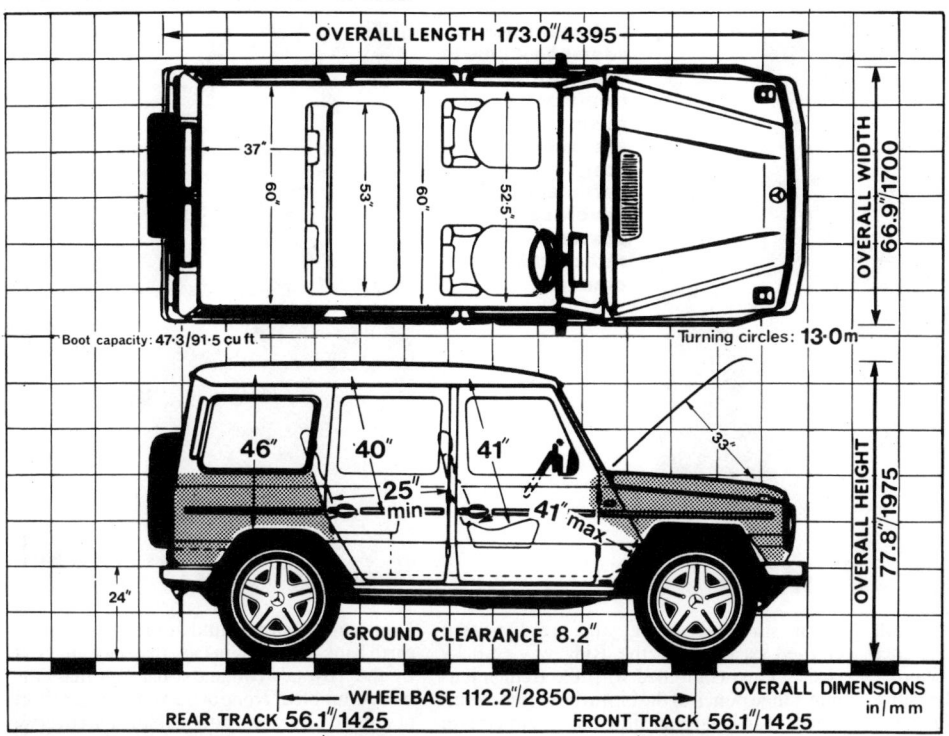

PERFORMANCE

MAXIMUM SPEEDS

Gear	mph	km/h	rpm
Top (mean)	84	135	4100
(best)	84	135	4100
4th	82	132	5000
3rd	61	98	5100
2nd	39	63	5200
1st	22	35	5200

ACCELERATION FROM REST

True mph	Time (secs)	Speedo mph
30	6.3	34
40	10.9	44
50	16.1	55
60	25.4	68
70	40.2	78
80	—	89

Standing ¼ mile 23.4secs, 58mph
Standing km 43.6secs, 71mph
30-70mph through gears 33.9secs

ACCELERATION IN EACH GEAR

mph	Top	4th	3rd	2nd
10-30	—	13.8	9.9	5.6
20-40	20.9	13.9	9.5	—
30-50	23.4	14.7	10.2	—
40-60	27.7	17.2	—	—
50-70	42.5	24.2	—	—

FUEL CONSUMPTION
Overall mpg 17.1 (16.5 litres/100km)
Touring mpg* 19.8 (14.2 litres/100km)
Government tests 18.8mpg (urban)
25.9mpg (steady 56mph)
17.7mpg (steady 75mph)
Fuel grade diesel
Tank capacity 20.9galls (95 litres)
Max range* 413 miles

*Based on Government fuel economy figures: 50 per cent of urban cycle, 25 per cent each of 56/75mph consumptions.

BRAKING
Fade (from 58mph in neutral)
Pedal load (lb) for 0.5g stops

start/end		start/end	
1	50-50	6	55-55
2	55-50	7	60-62
3	50-60	8	60-60
4	50-62	9	55-53
5	55-57	10	50-53

Response (from 30mph in neutral)

Load	g	Distance
10lb	0.07	430ft
20lb	0.24	125ft
30lb	0.45	67ft
40lb	0.45	67ft

WEIGHT
Kerb 5051lb/2293kg
Distribution % F/R 51/49
Test 5411lb/2456kg
Max payload 1168lb/530kg
Max towing weight 6497lb/2950kg

TEST CONDITIONS
Wind 4mph
Temperature 5deg C (41deg F)
Barometer 1001mbar
Surface damp asphalt/concrete
Test distance 950 miles

Figures taken at 2400 miles by our own staff at the Lotus Group proving ground, Millbrook.

All *Autocar & Motor* test results are subject to world copyright and may not be reproduced without the Editor's written permission.

SPECIFICATION

ENGINE
Longitudinal, front, permanent four-wheel drive.
Capacity 2996cc, 6 cylinders in line.
Bore 87.0mm, **stroke** 84.0mm.
Compression ratio 22.0 to 1.
Head/block cast iron/cast iron.
Valve gear sohc, 2 valves per cylinder.
Fuelling indirect diesel injection.
Max power 113bhp (PS-DIN) (83kW ISO) at 4600rpm. **Max torque** 141lb ft (191Nm) at 2900rpm.

TRANSMISSION
5-speed manual.

Gear	Ratio	mph/1000rpm
Top	0.80	20.5
4th	1.00	16.4
3rd	1.37	12.0
2nd	2.18	7.5
1st	3.86	4.2

Final drive ratio 5.29 to 1.
Front, centre and rear switchable diff locks.

SUSPENSION
Front, live axle, radius arms, coil springs, telescopic dampers, anti-roll bar.
Rear live axle, radius arms, coil springs, telescopic dampers, anti-roll bar.

STEERING
Recirculating ball, power assisted, 3.8 turns lock to lock.

BRAKES
Front 11.9ins (303mm) dia disc.
Rear 11.9ins (303mm) dia drums.
Anti-lock standard.

WHEELS AND TYRES
Cast alloy 7ins rims. Goodyear Wrangler 255/75R15 tyres.

SOLD IN THE UK BY
Mercedes-Benz UK Ltd
Mercedes-Benz Centre
Tongwell, Milton Keynes
MK15 8BA
Tel: 0908 668899

COSTS

Total (in UK) £31,890
Options fitted to test car:
Wide wheels and wheel arches £700.01
Electric tilting/sliding sunroof £997.00
Electric aerial and 4 speakers £384.00
Outside temperature gauge £115.00
Blaupunkt radio-cassette (est) £350.00
Total as tested £34,436.01
Delivery, road tax, plates £315
On the road price £34,751.01

SERVICE
Major service 12,000 miles. Service time 2.5 hrs. Intermediate service 6000 miles. Service time 1.8 hrs.

PARTS COST (inc VAT)
Oil filter £9.78
Air filter £18.24
Brake pads (2 wheels) front £70.47
Brake pads (2 wheels) rear £9.32
Exhaust complete £757.47
Tyre — each (typical) £152.37
Windscreen £55.12
Headlamp unit £51.18
Front wing £171.29
Rear bumper £379.53

WARRANTY
12 months/unlimited mileage, 1 year against paint defects, 12 months breakdown recovery

EQUIPMENT

Anti-lock brakes	●
Self-levelling suspension	●
Alloy wheels	●
Auto gearbox	£1370
Power-assisted steering	●
Locking differentials	●
Steering rake adjustment	●
Seat height adjustment	●
Electric seat adjustment	●
Lumbar adjustment	●
Head restraints	●
Intermittent wipe (variable)	●
Heated seats	£359
Leather trim	£2285
Air conditioning	£2578
Cruise control	●
Radio/cassette player	DO
Electric aerial	£206
Speakers	£178
Electric windows F/R	●
Central locking	●
Front fog/driving lamps	●
Headlamp wash	●
Electric tilt/slide	£997
Metallic paint	£750

● Standard
— Not available
DO Dealer option

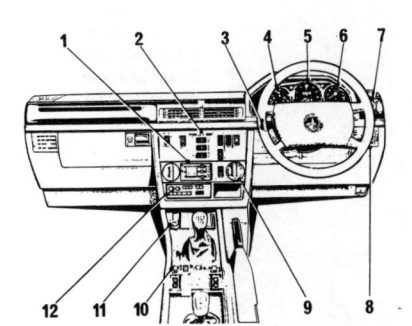

1 Heating and ventilation controls, **2** Switches for rear screen wash/wipe/differential locks/hazard/windows and ABS, **3** Indicators and wash/wipe stalk, **4** Temperature/fuel and oil pressure gauges, **5** Speedometer, **6** Tachometer and clock, **7** Lights, **8** Combination switch, **9** Fan, **10** Window switches, **11** Cigar lighter, **12** Radio/cassette

OFF-ROADERS

G-Wagen interior seems best able to cope with harsh treatment. Build quality and finish are excellent. Cargo area is sufficient but third row of seats are a cost option rather than standard

◀ the Ridgeway would provide us with a stiff test of both traction and ground clearance thanks to several weeks of heavy rain before.

Heading west from Streatley in Berkshire there was certainly no shortage of water on the track, and heavy showers throughout the day were to keep the chalk well and truly doused. We had elected not to lower the tyre pressure on the vehicles in an effort to expose their limits sooner, so with low range engaged and centre diffs locked we headed off-tarmac.

It didn't take long to appreciate that the new Landcruiser is in a completely different league to its dismal predecessor. Supple suspension that's progressively cushioned at the limits of its travel delivers a very good ride off-road, helped of course by its long wheelbase and sheer mass. With no low-slung leaf springs to worry about, ground clearance is very much better, although the running boards are prone to contact with the sides of deep ruts. So wide are its 265/75 Dunlops that they wedge themselves between the walls of deep ruts, and while that improves traction it means the Toyota is reluctant to leave a set of ruts.

On smoother going, the tread pattern clogged easily with the Ridgeway's sticky chalk. This, combined with the tremendous mid-range punch of the big turbo diesel, meant that even in low-range fifth only millimetres of throttle travel could be used before the Toyota started to wheelspin away all its traction. To some extent this was a problem that afflicted all the vehicles, and despite its lack of urge, the G-Wagen's wide Goodyear Wranglers were also prone to losing grip and were reluctant to leave ruts. On narrower 205 R16 Wranglers the Discovery was better able to cut through the sticky surface and find traction, but though they offer greater steering precision on the road the Goodyears are not a match for the Discovery's alternative-fit Michelin XM+S 244 tyres on mud.

Despite this, all three vehicles came through the worst the Ridgeway had to offer with flying colours. Even where differentials were ploughing their own furrows between the ruts, the Ridgeway's relatively mild gradients allowed the trio to keep moving, and we had to resort to the tow rope just once when the Landcruiser ran aground at the crest of a soft earth bank — so soft, in fact, that once attacked by the Toyota's two-and-a-half-ton mass the bank crumbled. None of the vehicles was short of ground clearance, although the Discovery's towbar did its usual imitation of a ground anchor at times.

From deep ruts and sticky mud we trekked to the steep, rocky slopes of the Brecon Beacons for day two of our off-road showdown. Freezing temperatures and driving sleet would make the rock steps and off-camber mountain-side tracks a real test of wet grip, articulation and sheer pulling power. First up in the Discovery, a combination of the Tdi engine's fine bottom-end punch and a significantly lower low range than the other two meant the steepest climb could be tackled in third gear. With no cross-axle diff locks the Land Rover product relies on fine articulation to keep all four tyres firmly on the ground, and by threading a careful line between the sharpest-edged rocks it made light of the climb.

Firmer sprung than either the Landcruiser or Discovery and with anti-roll bars to ▶

Discovery's blue trim is a refreshing departure but doesn't hide dirt. Cargo area is versatile and storage space is plentiful. Casual rear seats fold away neatly. Grab handles are everywhere

TOYOTA LANDCRUISER VX

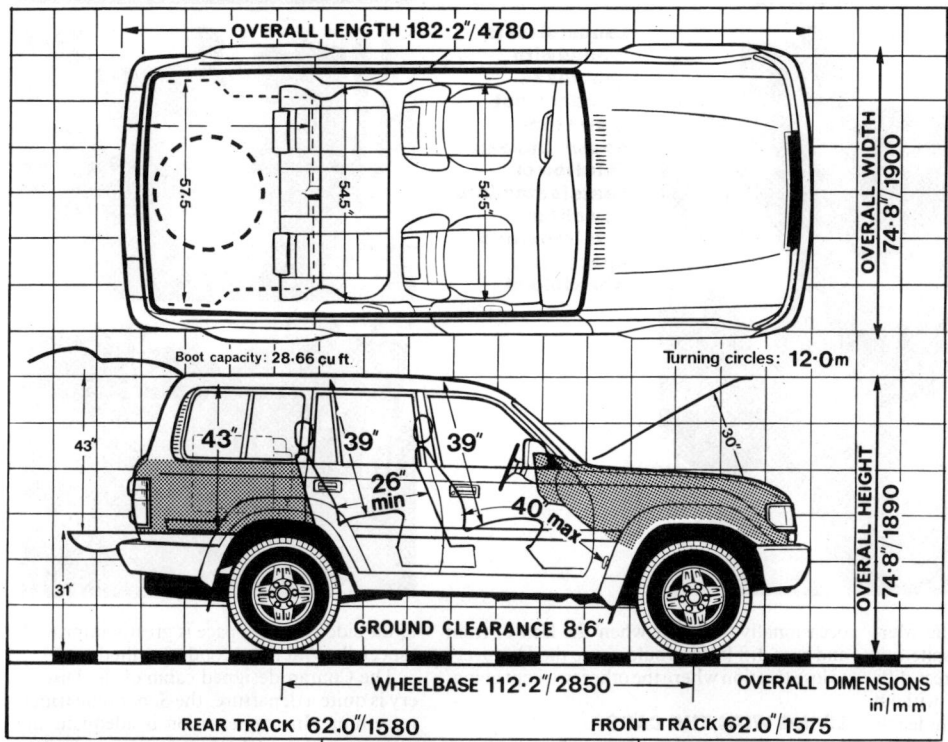

PERFORMANCE

MAXIMUM SPEEDS
Gear	mph	km/h	rpm
Top (mean)	103	166	3870
(best)	104	167	3910
4th	96	155	4100
3rd	67	108	4300
2nd	44	71	4300
1st	24	39	4300

ACCELERATION FROM REST
True mph	Time (secs)	Speedo mph
30	4.0	33
40	5.9	43
50	9.2	54
60	12.6	65
70	17.9	76
80	24.1	86
90	34.9	97
100	—	108

Standing ¼ mile 19.2secs, 72mph
Standing km 35.0secs, 91mph
30-70mph through gears 13.9secs

ACCELERATION IN EACH GEAR
mph	Top	4th	3rd	2nd
10-30	—	12.3	7.2	4.1
20-40	14.0	11.1	5.9	3.8
30-50	11.8	9.1	5.4	—
40-60	10.4	8.4	6.1	—
50-70	11.5	9.3	—	—
60-80	13.6	11.0	—	—
70-90	17.6	16.0	—	—

FUEL CONSUMPTION
Overall mpg 19.8 (14.3 litres/100km)
Touring mpg* 22.9 (12.3 litres/100km)
Govt tests mpg: 23.3mpg (urban)
29.1mpg (steady 56mph)
18.3mpg (steady 75mph)
Fuel grade diesel
Tank capacity 21galls (95 litres)
Max range* 479 miles
*Based on Government fuel economy figures: 50 per cent of urban cycle, 25 per cent each of 56/75mph consumptions.

BRAKING
Fade (from 72mph in neutral)
Pedal load (lb) for 0.5g stops
start/end		start/end	
1	25-22	6	30-40
2	25-25	7	32-38
3	25-22	8	34-36
4	30-32	9	32-35
5	30-35	10	35-35

Response (from 30mph in neutral)
Load	g	Distance
10lb	0.20	150ft
20lb	0.40	75ft
30lb	0.55	55ft
40lb	0.72	42ft
50lb	0.95	32ft
60lb	1.10	27ft

WEIGHT
Kerb 4971lb/2257kg
Distribution % F/R 54/46
Test 5332lb/2421kg
Max payload 1222lb/555kg
Max towing weight 7709lb/3500kg

TEST CONDITIONS
Wind 12mph
Temperature 22deg C (72deg F)
Barometer 1022mbar
Surface dry asphalt/concrete
Test distance 1200 miles

Figures taken at 6000 miles by our own staff at the Lotus Group proving ground, Millbrook.
All *Autocar & Motor* test results are subject to world copyright and may not be reproduced without the Editor's written permission.

SPECIFICATION

ENGINE
Longitudinal, front, permanent four-wheel drive.
Capacity 4164cc, 6 cylinders in line.
Bore 94mm, **stroke** 100mm.
Compression ratio 18.6 to 1.
Head/block al alloy/cast iron.
Valve gear sohc, 2 valves per cylinder.
Fuelling direct diesel injection, turbocharger.
Max power 165bhp (PS-DIN) (121kW ISO) at 3600rpm. **Max torque** 265lb ft (195Nm) at 1800rpm.

TRANSMISSION
5-speed manual.
Gear	Ratio	mph/1000rpm
Top	0.881	26.6
4th	1.000	23.4
3rd	1.490	15.7
2nd	2.294	10.2
1st	4.081	5.7

Final drive ratio 3.727 to 1. Diff locks front, centre and rear.

SUSPENSION
Front live axle, radius arms, coil springs, telescopic dampers, anti-roll bar.
Rear live axle, radius arms, coil springs, telescopic dampers, anti-roll bar.

STEERING
Recirculating ball, power assisted, 3.8 turns lock to lock.

BRAKES
Front 11.2ins (286mm) dia disc.
Rear 12.3ins (312mm) dia disc.

WHEELS AND TYRES
Cast alloy 7ins rims.
Dunlop SP RV Major 265/75 SR15 tyres.

SOLD IN THE UK BY
Toyota (GB) Ltd
The Quadrangle
Redhill, Surrey RH1 1PX
Tel: 0737 768585

COSTS
Total (in UK) £27,430.76
Options fitted to test car:
Metallic paint £199.00
Total as tested £27,629.76
Delivery, road tax, plates £415.00
On the road price £28,044.76

SERVICE
Major service 24,000 miles. Service time 2.8 hrs. Intermediate service 12,000 miles. Service time 2.5 hrs. Oil change 6000 miles. Service time 2.2 hrs.

PARTS COST (inc VAT)
Oil filter	£11.70
Air filter	£20.45
Brake pads (2 wheels) front	£27.89
Brake pads (2 wheels) rear	£35.76
Exhaust complete	£184.44
Tyre – each (typical)	£145.81
Windscreen	£90.62
Headlamp unit	£52.94
Front wing	£165.58
Rear bumper	£123.33

WARRANTY
36 months/60,000 miles unlimited mileage, 6 years anti-corrosion, 3 years against paint defects, 12 months recovery.

EQUIPMENT
Anti-lock brakes	—
Self-levelling suspension	—
Alloy wheels	●
Auto gearbox	£1125
Power assisted steering	●
Locking differentials	●
Steering rake/adjustment	●
Seat height adjustment	●
Electric seat adjustment	●
Lumbar adjustment	●
Electric door mirrors	●
Intermittent wipe (variable)	●
Heated seats	●
Leather trim	●
Air conditioning	£1604
Cruise control	—
Radio/cassette player	●
Electrical aerial	—
5 speakers	●
Electric windows F/R	●
Central locking	●
Front fog/driving lamps	—
Headlamp wash	●
Electric tilt/slide sunroof	£199
Metallic paint	£199

● Standard
— Not available

1 Radio/cassette player, 2 Sub fuel tank/power antenna/and rear screen switches and diff lock button, 3 Rear screen wash/wipe, 4 Hazard warning, 5 Oil pressure/engine temperature gauges, 6 Tachometer, 7 Speedometer, 8 Voltmeter and fuel gauge, 9 Indicator and lights, 10 Bonnet release and fuel filler opening, 11 Steering rack adjuster, 12 Front/rear differential lock, 13 Heating and ventilation controls, 14 Panel rheostat, 15 Cigar lighter

OFF-ROADERS

Landcruiser is loaded with goodies as standard, but brown trim doesn't appeal. Third row of seats is standard as well but cuts into cargo space even when folded

inhibit single-wheel movement, the Mercedes lifted an occasional wheel despite our best efforts, yet with just centre and rear diffs locked it too reached the summit without showing any signs of losing traction. The feeble mid-range response of the 3-litre six and the smallest gap between low and high ratios of the group did mean that the G-Wagen needed second rather than third to make it up the hill.

Its terrific low-speed response meant the Landcruiser had plenty of torque in reserve for the climb, and in truth it kept its wheels on the ground rather better than the G-Wagen. Still we kept the rear diff locked (the centre is automatically locked in low range), and counting against the Toyota on particularly tight sections were its bulk and poor turning circle. The sidewalls of its very wide Dunlops were harder to keep away from potentially damaging rock outcrops, though all three off-roaders escaped without a single puncture.

Locking the front diff on either the Mercedes or Toyota inevitably makes the steering very cumbersome, but rearward weight transfer means it really isn't necessary on most climbs. Of more concern is the Toyota's kangaroo-like throttle response, making picking a path through the boulders a very tricky operation even in a high gear. The same problem can cause surging when coming down a steep slope in low-range first, despite a high tickover that means ultimate engine braking is not as great as it might be. Easily the most controllable down a steep hill is the Discovery, which combines terrific engine braking from the Tdi engine with by far the lowest low-range first gear.

Make no mistake, these vehicles are all excellent off-roaders. The Mercedes was very able in its previous part-time four-wheel-drive incarnation, and the new G-Wagen is tough, capable and let down only by its mediocre performance in diesel guise. Toyota's new Landcruiser is a far better tool than its predecessor and its diesel engine delivers terrific performance on the rough. As the lightest vehicle here, the Discovery makes very good use of its punchy Tdi engine, and with no anti-roll bars its Range Rover-derived chassis has the articulation and control to deliver as much traction as the other two with just a single, central diff lock.

Nevertheless, instances will inevitably — if occasionally — occur when the lack of front and rear diff locks would cause the Discovery to lose traction where the other two kept going.

LIVING WITH THE CARS

In many ways these vehicles can be seen as over-sized estate cars charged with ferrying the kids, Fido and weekly provisions from A to B with monotonous regularity.

Naturally, space and cabin robustness are key factors in the long run and, to this end, the Mercedes is the only one that really looks like it's built to endure a life of human and canine abuse. Predominantly constructed of two-tone moulded grey plastic, the German car's interior now owes little to its military origins. As usual, the Mercedes' seats are firmly sprung and feel like they have been sensibly uprated from the norm to iron out the harshest of bumps.

In typical Mercedes fashion, the instruments are clear and uncluttered, if a little small, and the main binnacle differs little from that found in the mid-range saloons. Prominent among the sensibly placed switchgear are the three diff-lock rocker switches, which can only be engaged in the correct sequence.

All-round visibility is good and the large door mirrors are electrically operated. The rear seats split asymmetrically and can be folded for increased luggage space. A cost option for the long-wheelbase G-Wagen is extra rear seating to provide full seven-seater accommodation.

Our test car had an electric tilt and slide sunroof, outside temperature control, wide alloy wheels and wheel arches, an electric aerial, four speakers and a Blaupunkt Cambridge radio/cassette. This adds £2546 to the price.

As you would expect from the Japanese, the Landcruiser comes loaded with all the goodies you could want and more. Precise adjustment of the driving position is possible due to an impressive array of seat movements and rake adjustment on the steering wheel. Less appealing is the brown cloth and carpet with which the Toyota is trimmed. The instrumentation fails to provide the level of clarity achieved in the Mercedes but is still acceptable.

With its third row of seats installed, the Landcruiser's luggage space all but disappears. However, with the seats folded neatly to the side, storage space is greatly improved, especially with the second row tilted up.

The Conran-designed cabin of the Discovery is quite a departure: the Sonar blue trim is refreshing. Instrumentation is adequate and no more, but most disappointing is the heating and ventilation system, which takes an age to provide heat when starting up and is hopeless at demisting. Our test car had a number of options fitted — two manual tilt sunroofs, a high spec audio system, a towing pack and Micatallic paint totalling £1366.

For storage space, the Discovery wins hands down. Wherever possible its designers have put in handy oddment bins, non-slip surfaces and elasticated string bags. A nice touch too are the grab handles to be found everywhere. The Land Rover has split rear seats that tilt forward to increase storage space, and neatly folded to the sides and positioned to the rear are two casual seats complete with lap belts. Of the three, the Discovery is the most versatile and makes best use of available space.

The Land Rover took the fuel honours, returning an overall figure of 23.3mpg, with the Toyota second (19.8mpg) and the Mercedes last with a disappointing 17.1mpg.

THE VERDICT

In 300GD long-wheelbase form at least, the G-Wagen is a frustratingly flawed machine. We wouldn't seek to dispute its engine refinement, build integrity, finish or sheer go-anywhere tenacity but we do have a problem with its desperate lack of performance and feeble brakes. In the light of its high price, the G-Wagen most emphatically collects the wooden spoon in this company.

Much more like it is Toyota's imposing and impressive Landcruiser. It's big, well built, generously equipped, refined and very capable — both on and off road. Putting price to one side for the moment, it's the best 4×4 here.

By the narrowest of margins, however, it's the Discovery Tdi that takes top slot. On the road it can't compete with the Toyota but it's far from disgraced and, when the going gets tough, it just keeps going. The build and finish aren't all they might be but, that said, it's much cheaper than its rivals. As such, it continues to be a bargain. We'd happily take the Discovery and keep the change. ∎

TRIED

LAND ROVER DISCOVERY V8i

Launched at the end of 1989, the Discovery was the first new model from Land Rover for 19 years. Now the carburettor-fed 3.5-litre V8 has recieved fuel injection. Andrew English tries it.

IT MUST have been the world's worst kept secret, except Land Rover wasn't admitting anything. When the Discovery was first shown at the end of 1989, everyone knew what it would look like, what the engine options were and how it would be sold. In spite of this, most of the press caught Discovery fever; one magazine devoted over 30 pages to it, including two front covers.

Were we all conned? Had the 40-year-old manufacturer played the press and the public into believing that the Discovery was something more than a rebodied Range Rover? After all, we all knew Land Rover was losing market share to the Japanese leisure off-road vehicles, and it had to have a vehicle to go between the Range Rover and the newly renamed Defender.

Probably not, but the Discovery inauguration was nothing short of a *tour-de-force* for the Solihull company. It couldn't have come at a better time. The demoralised work force had crawled back to work after a fruitless six-week strike. The Ministry of Defence had sent back thousands of military Land Rover diesel engines when faults were found and sales had really started to fall against Japanese competitors. A new vehicle and a new engine (the turbo diesel TDi) provided the necessary tonic.

One year on, the Discovery has been subtly modified with cast-offs from the Range Rover. It has moved further upmarket, with the version tested here costing a basic £20,470. Range Rover and Discovery production in the period from January to September 1990 was 60.1percent up on 1989's, which goes against a fairly gloomy industry trend. So what is all this Discovery stuff? Is it as good as the sales figures say it is? If not, why is everyone buying it?

The Discovery driven here in the latest specification V8i, with five-door bodywork and the 3.5-litre uncatalysed fuel-injected V8 that was fitted to the Range Rover until US dealers asked for, and received, a 3.9-litre version.

The chassis is the same as the Range Rover's with the exception of the Boge self-levelling system. Coil-sprung beam axles at both ends are located with radius arms and a Panhard rod at the front and trailing links and an A-frame at the rear. The five-speed Rover gearbox handles the engine's torque (212lb ft), feeding it into a chain-driven two-speed transfer box with a manual differential lock. The Range Rover gets a viscous coupling here, as well as ABS.

The body is aluminium with the exception of the complicated steel roof pressing. It is slightly bigger than the Range Rover and seats seven.

Outside designers (including Conran) were used to develop the interior, and it shows. Despite the plethora of pockets, cubbies and nets, there is a severe lack of usable storage space inside. The centre console ejects its contents when the vehicle tackles corners with a degree of gusto, and the map pockets in the door are completely inaccessible when the doors are closed.

To keep the price down, the Discovery interior is all plastic. No amount of dimpled surfaces nor weird-shaped grab handles can hide its basically cheap nature. It looks straight out of Play School; wipe-clean, and easy for small hands.

Complaints include impossibly complicated heater and optional air conditioning controls, awkwardly located door handles, poor dashboard binnacle switches and dim headlamps (no they weren't covered in mud!). We like the decent sized floor mats, and the Austin Rover parts-bin steering column stalks are probably the best in the business.

The bodywork is surprisingly well finished, with properly narrow shut lines, undimpled flat areas and a general air of high quality. It has also been described in the past as ugly, and we see no reason to disagree with this now.

Performance is surprisingly good for this class of vehicle, and the poor aerodynamics only start to be felt above 50mph. Initial acceleration is brisk, and there is lots of torque for off-road work. Fully-laden motorway bashes and some off-road adventures resulted in a fuel consumption of 14.6mpg. Don't expect much more than this, even if you are light-footed.

I have always liked the Rover-sourced 77mm gearbox, but then I drive a Triumph TR4, a euphemism for notchy gearboxes. With time, the corners wear off the gearchange and fast shifts can eventually be made. The transfer box is not such good news, and engaging the stubby little lever demands careful feeding in with the clutch. Likewise the differential lock can be tricky to engage, and both hands are recommended.

Handling in an off-road vehicle can never be described as brilliant; the Discovery is better than most. It rolls slightly less, and the steering at least *tries* to communicate. We just didn't understand the language. Unfortunately, like in most of its contemporaries, the driver is only part of a democratic process in deciding vehicle direction, with tyre and vehicle roll, slip angles and beam axle movement all having their vote. The Discovery is at its unhappiest on twisting B-roads with irregular radius curves. If you can see the corner entry and exit, and set the vehicle up, the roll is tolerable. With blind corners, the soft suspension makes the necessary rapid changes of direction a bit like piloting a ship in a gale; pass the sea-sickness pills.

Off road, the Land Rover chassis is second to none. Excellent wheel travel and articulation ensure that progress can be made in the very worst of conditions. You can still get the vehicle stuck if you try, and the lack of limited-slip or locking differentials in either axle means that the Discovery can have problems if a front and rear wheel start to spin simultaneously.

Lots of people laugh at off-road leisure vehicles, saying they do little more than lift a wheel onto the pavement in Chelsea. Quite apart from the fact that a lot of Discoveries are doing the arduous towing or off-road jobs for which they were designed, the vehicle is uniquely suited to the urban way of life as well. The high driving position, good acceleration and light steering make it easy to drive around town. There is plenty of room for shopping, children and stuff. It is easy to park with its good visibility and the excellent suspension soaks up the worst of the inner-city third world roads. Rapid weekend motorway jaunts are entirely within its capabilities, and it is different. Who's laughing now?

We like the Discovery. It's one of the better off roaders around, and is very practical. The Range Rover is the better and classier vehicle, but it costs too much. If hot-shoe motoring is not for you, and you don't do too much minor road driving, the Discovery provides an alternative.

Price: £20,470. Verdict: Ugly but very sound off-road monster. Shame about the plastic interior, but a good (and probably more practical) alternative to a second-hand Range Rover.

The Discovery stands up in the role of long-haul tourer as well as peerless off-roader, as a 2500-mile round trip to Poland proved (left). But the Land Rover still shows its true mettle in agricultural conditions. Dash is concise, but partially obscured. Switchgear shows signs of cost cutting. Heating and ventilation are pathetic

LONG TERM Land Rover Discovery Tdi

With Discovery, Land Rover has shaken off its agricultural associations and is chasing the Japanese-led recreational market with a vengeance. After 21,000 miles Martin Vincent is convinced by the design but doubtful about build quality

LAND ROVERS ARE TOUGH, AREN'T they? Faced with a seemingly impossible ascent of rocks and ditches, yes they are — the company has built its reputation on the fact. The Discovery, the company's bold and successful initiative to take on the Japanese-dominated 'recreational' four-wheel drive class, is no different from any other Land Rover in this respect. Great news if you happen to live at 5000ft with only that rocky road to your front door.

We don't, and probably 99 per cent of Discovery owners don't either. Which isn't to say we shouldn't have bought one. We wanted a Discovery for its go anywhere ability, sure, but we — like most owners, we suspect — wanted it also for its towing ability, cabin room, seven seats, big glass area and high seating position that you never knew could mean so much until you lived with them. The reality of the market Land Rover has entered with the Discovery is one of commuter crawling and motorway cruising. Ignoble duty for one so tough.

Tough? No, faced with the ultimate test — with the family aboard to Sainsbury's for the weekend shop — the Discovery is actually a bit of a wimp.

The harsh verdict after a year and 21,000 miles with our three-door TDi is that were it a family hatchback with no other side to its character, we would probably have sent it back. The niggling faults, the ongoing feeling of fragility and cheapness inside, the two roadside breakdowns and the questions we have had to ask over the longevity of some major components are all out of step with the level of build quality of even the lowest-priced hatchbacks.

The soft verdict, and the one we incline to more because, damn it, we like the Discovery, is that here is the annoying but far from disastrous price you pay for enjoying one of the best conceived, designed and priced all-rounders on the market.

Three months of twiddling thumbs followed our initial order for G373 EPC. It finally appeared on our doorstep on 30 March 1990, resplendent (in an orange peel sort of way) in micametallic arken grey coachwork. Then, the base price was £15,750, but once the seven-seat option, towing pack, metallic paint and the 'value pack' were added, plus delivery and road tax, the final bill came to £18,168.25. The 'value pack' consists of central locking, electric front windows, electrically operated and heated door mirrors, headlamp power wash, twin sunroofs, load-space cover and roof bars.

A year on, it's a lot easier to obtain a Discovery, although that three month wait still applies if you're fussy about colour and specification. It doesn't seem to have put people off too much: in 1990 the Discovery outsold its nearest rival, the Mitsubishi Shogun, by two to one. And with the latest five-door and injected V8 engined versions it seems set to retain the top slot for a while yet.

To buy a three-door Discovery TDi with a similar specification to ours, the going rate today is £20,036 (£17,300 current basic price). Allowing for high mileage, a dealer would ▶

WHAT IT HAS COST
Price new £15,750 (with options, £18,168)
Price now £17,300 (with options, £20,036 approx)
Estimated trade in value £14,500

FUEL/OIL (Cost for 21,456 miles)
3577litres (788gals) of diesel £1497.20
1.0 litre of oil £2.40
Total £1499.60

TYRES
Cost (each, est) £88.94
Life Front - 65% worn £115.62
Rear - 40% worn £71.15

SERVICE AND REPAIR RECORD
Faults on delivery: None

SERVICE
Mileage	Date	Cost
6,000	7.6.90	£92.68
12,000	26.10.90	£296.32
18,000	8.1.91	£180.65

REPAIRS
Replace offside door mirror at 6000 mile service

EXTRAORDINARY ITEMS (fixed under warranty)
Ignition switch faulty, broken tailgate latch, broken sunroof catch, rear window catch fell off, tailgate dropped on hinges, loose fuel injector pipes.

ANNUAL STANDING COSTS
Road tax £100.00
Insurance premium* £240
* to put cars on equal footing for insurance costs, the figure given is for a typical quotation for a good-risk driver, with a clean record, and car garaged in Oxfordshire, a middle-risk area. Full no claims discount has been deducted. Source: Quotel Motor Insurance Service

DEPRECIATION
(12 months estimated) £3668

TOTAL RUNNING COSTS
(12 months ex depreciation) £2596.02

COSTS PER MILE
(Including Road Tax, insurance) 12.1p
(Including Road Tax, insurance and depreciation) 29.2p

SPECIFICATION
LAYOUT
Longitudinal, front, permanent four-wheel drive

ENGINE
Capacity 2495cc
Bore 90.5mm, **Stroke** 97.0mm
Compression ratio 19.5:1
Head/block al. alloy/cast iron
Valve gear ohv, 2 valves per cylinder
Fuel Direct diesel injection, Garrett T25 turbocharger
Max power 111bhp (PS-DIN) (83kW ISO) at 4000rpm.
Max torque 195lb ft (265 NM) at 1800rpm

TRANSMISSION
five-speed manual

Gear	Ratio	mph/1000rpm
Top	0.770	25.1
4th	1.000	19.3
3rd	1.397	13.9
2nd	2.132	9.1
1st	3.692	5.2

Final drive ratio 3.538:1. Transfer ratio: high 1.222, low 3.320, lockable centre diff.

SUSPENSION
Front, live axle, radius arms, Panhard rod, coil springs, telescopic dampers.
Rear, live axle, radius arms, upper A-frame, coil springs, telescopic dampers.

STEERING
Recirculating ball, power assisted, 3.8 turns lock-to-lock.

BRAKES
Front 11.8ins (299mm) dia. discs **Rear** 11.4ins (290mm) dia. discs.

WHEELS AND TYRES
Pressed steel, 7ins rims. Goodyear Wrangler 205R16 tyres

SOLD IN THE UK BY
Land Rover Ltd., Lode Lane, Solihull, W. Midlands. B92 8NW. Tel: 021 722 2424

PERFORMANCE
MAXIMUM SPEEDS

Gear	LT mph	LT rpm	RT mph	RT rpm
Top (mean)	91	3620	92	3650
(best)	92	3650	92	3650
4th	84	4350	84	4350
3rd	63	4500	63	4500
2nd	41	4500	41	4500
1st	23	4500	23	4500

Standing quarter mile: LT 20.5secs/65mph
RT 20.5secs/63mph
Standing km: LT 38.5secs/81mph
RT 38.8secs/81mph
30-70 thro' gears: LT 19.9secs RT 20.6secs

ACCELERATION
From rest

True mph	LT Time secs	RT Time secs	Speedo mph LT
30	4.5	4.7	31
40	7.5	7.6	41
50	11.2	11.5	51
60	16.5	17.1	62
70	24.4	25.1	72
80	37.3	37.1	83

ACCELERATION IN EACH GEAR

mph	top LT	4th LT	3rd LT	2nd LT
10-30	—	14.0	8.0	4.7
20-40	19.9	10.0	6.2	5.1
30-50	14.6	8.9	7.0	—
40-60	15.2	10.2	9.7	—
50-70	18.7	13.6	—	—
60-80	28.2	22.1	—	—

LT denotes long term car; **RT** denotes performance figures for Discovery TDi tested in *Autocar & Motor* on 15th November 1989

FUEL CONSUMPTION
Average LT mpg 27.2
Best/Worst LT mpg 31.9/20.6
Average RT mpg 23.9

◀ offer about £14,000 to £15,000 for our year-old example. Were we to sell it privately, we'd be looking to get £16,500. Compared with the executive sector, where a 40 to 50 per cent first year drop is common, it's a good result.

And in several key ways living with our Discovery as the miles pile on has been good, too. The short travel gearchange has loosened up and is now pleasantly precise and meaty, rather than stiff and awkward as at first. The only real gripe now is the ease with which reverse can be engaged in place of first. The engine gets stronger all the time; now that it's nicely loosened up it's quick enough to make newer turbo diesel Discoveries seem sluggish. Our figures, taken at 21,000 miles, show significant gains on the road test TDi three-door tested with 4100 miles on the odometer. And it is dramatically quicker than the five-door TDi tested recently.

We're delighted with the economy, too. A 2500-mile trip to Poland produced an average of 27mpg. Country roads see this figure rise beyond 32mpg, but city traffic causes the economy to fall to about 21mpg. Overall it's a very acceptable result from a heavy, unaerodynamic seven-seater.

The engine is flexible, too, and has good response for a diesel, but we do have reservations about its refinement. On starting up from cold, the direct-injection Land Rover turbo diesel clatters into life, accompanied by vibrations that set the entire steering column and instrument housing shaking. It smoothes out within a few moments, but there's still an underlying vibration, which, combined with the mechanical clamour, can be wearying on a long trip.

And the engine is just too noisy. With such a weedy radio — one of a number of conspicuously cheap components — even turning up the volume doesn't hide the cacophony. The thief who broke a side window and attempted to steal the radio realised his error in time: the radio was left half removed but not working. He unwittingly did us a favour. A more powerful Philips pull-out unit with RDS has now taken its place.

Despite its off-roading ability, the Discovery has proved to be a more than capable all-rounder on tarmac, in the manner of bigger brother Range Rover. It takes motorways in its stride and, around town or in the country, there is little to disadvantage it from more conventional transport — other than a comparative lack of overtaking urge. Plan your move ahead, though, and you can make good times.

With power steering and the high driving position, parking is easy enough, but the poor access to the rear seats afforded by the three-door body makes off-loading or loading passengers a nuisance. Getting kids and/or baby seats aboard and strapped in is a good enough reason by itself to upgrade to the five-door, which wasn't available when we bought our Discovery. Children are hapy to sit on the two neatly folding seats in the rear luggage area, however, while wherever you sit you can't help but like the terrific view.

The heavy clutch, with plenty of room in the footwell for size 11 wellies, is appropriate to the car's character, but we are not so sure about the equally heavy transmission handbrake that never seems totally trustworthy.

The TDi stops and grips securely and, in snow and heavy rain, shows its Land Rover pedigree to great effect. Only in the vague steering, and the alarming body roll when attacking corners, does the Discovery make it plain that this is an off-roader first and foremost. At first the steering seems disconnected from the wheels, almost as if it has a delayed action effect. But with perseverance you can learn to live with it.

With long travel, pliant coil-sprung suspension and firmly supportive seats, we have few complaints concerning comfort. Sharp-edged imperfections cause a judder through the structure, but there's little else that will shake its composure.

Off-road the Discovery shows its true mettle. Whether tackling a mildly difficult green road with the family and a wadge of Ordnance Survey maps or exploring the limits of gradeability on a wet Welsh mountain, the Discovery never puts a foot wrong. I learnt just how deep its abilities run at Peter Morgan's Motor Safari off-road school in North Wales. You can only be amazed at the steep inclines it can tackle in expert hands, and how effectively the combination of low-range gearing and a diesel engine can tame seemingly impossible descents.

For anyone looking at this class of vehicle, the Discovery's cabin design is one of its greatest strengths. The first class interior from the Conran Design group, fresh ideas and thoughtful detailing abound in what is, by general 4wd standards, an attractive, airy and practical interior.

However, there is an insubstantial quality to many of the interior fittings and we were concerned initially at how they would stand up to everyday use. So far there have been no major failings, although the rear luggage cover, which never really fitted properly in the first place, is starting to fall apart. You must use it often, too, since with the lack of any lockable storages spaces it is the only place to leave items out of sight of thieving eyes. Another annoying thing is the door pockets, which you can only get at when the doors are open.

The cabin remains rattle free, but a 'zizzing' sound occasionally emanates from the depths of the facia, and the headlining seems unreasonably floppy. Also, the gear knob has worked loose. We remain unconvinced that this interior could withstand a boisterous family over a long period.

The minor switchgear shows signs of cost-cutting, with over-light and cheap-feeling stalk switches. The remaining dashboard switches are partially obscured by the chunky steering wheel.

Another area ripe for improvement is the prehistoric heating and ventilation system. The controls are confusing, awkward and crude, and air throughput is poor even when the heater eventually wakes up. Demisting is best achieved by opening a window.

So far, we've been left stranded at the roadside twice. The first serious malfunction occurred at 9300 miles, when the Discovery stubbornly refused to rumble into life. Our local dealer, Guy Salmon of Thames Ditton, promptly showed up with a loan car, and soon after the problem was traced to a faulty ignition switch.

The other immobilising episode occurred when one of the diesel injection return pipes came loose and sprayed fuel liberally around the engine bay. This was easier to fix, as the parts which worked loose came to rest on the engine block. It was a simple matter to re-attach the pipe and to tighten the three remaining ones, which were also loose. From correspondence we've received since, our TDi isn't the only Discovery to have suffered this problem.

Other annoying faults include a broken plastic sunroof catch, a side window catch that vibrated loose and finally fell off, and the rear door handle, which broke at its pivot. At 18,000 miles, the rear brake pads needed replacing, and the fronts will also need changing at the next service. Also, for some inexplicable reason, the headlights needed readjustment, as did the rear door, which had dropped on its hinges so it scraped the bumper whenever it was opened.

Cosmetically, the Discovery has coped

Discovery's excellent four-wheel drive system deals with all types of road surface with impunity. Conran interior gets top marks for design, but durability of materials is in doubt. Long term example comes with extra rear seats — making it a seven-seater — and rear load cover. Luggage space is simply gigantic

with a year's hard use better than expected. The paint always showed signs of poor application but it has retained its gloss, and the only trace of rust on the chassis is on the central cross-member where it was scraped during an off-road outing. The pressed steel wheels haven't survived so well — corrosion is beginning to appear through the thinly applied silver paint.

Under the bonnet, evidence of surface corrosion can be seen on the fuel filter housing, steering linkage and clutch master cylinder. Beneath the transmission casing, a patch of oil indicates there's a leak somewhere in the region of the gearbox or transfer box. It will be attended to at the next service, when the transmission idler gear chatter will be eradicated with the addition of extra O rings.

By the 18,000-mile service, a clonking from the drivetrain had reached worrying proportions. According to Guy Salmon's service technicians, our Discovery isn't any worse than normal for a four-wheel drive vehicle, which naturally has more driveline shunt than a conventional car. However, it wasn't always so harsh and noisy in its drive take-up, and neither were any of the other Discoveries we've tried recently. Ominously, the clonking seems to be getting worse.

Dealer attention is now imminent, and we will bring you up to date with all further developments, which, we sincerely trust, will put us on the road to a further rather less troublesome 21,000 miles. From the evidence of driving more recently built Discoveries (and ours is an early one, remember) build standards are on the up.

They need to be, for this new breed of customer Land Rover is after with the Discovery will surely find it impossible to downgrade expectations of quality and reliability, just as Land Rover cannot afford to downgrade off-road ability however few of its customers use it to the full. But our experience after 21,000 miles is that there needs to be a better balance between build and ability.

Rover's quality halo has a place above Solihull, but we're not sure it's there yet. ∎

DRIVELINES

Land Rover Discovery

The Discovery is the first all-new-from-the-ground-up model to be launched by Land Rover in nearly 20 years and with contemporary styling, a luxurious in-car feel and an outstanding chassis it is arguably the most impressive interpretation of Range Rover philosophy yet. Inevitably, it will be dubbed "the poor man's Range Rover".

Like the many varied Japanese offroaders to head our way in recent years, the Discovery's emphasis is well and truly on user-friendliness as much as ultimate offroad ability, and priced at $44,995 will pose a big sales threat to the more established makes in a sector of the market that has gradually been relinquished by the Range Rover, owing to its elevation into BMW and Mercedes-Benz territory.

For those desiring a Range Rover (and who doesn't from time to time?) but baulk at its artificially inflated price, at half the price the Discovery shines brightly as a perfect alternative. Even when tizzied up with each and every piece of optional equipment (there are 35 extras) Land Rover's latest still falls about $40,000 short of the top-line Range Rover. But it's on- and off-road ability is as good if not better than the boxy 20 year old from Solihull.

Designed to run on regular unleaded petrol, the Discovery's fuel-injected 3.5-litre V8 — essentially the same unit that powered earlier Range Rovers — pumps out 115 kW of power and 260 Nm of torque. It's a good engine with a nice strong feel from way down in the rev range; it will wind out without protest from under 20 km/h in fifth gear to a claimed maximum of 163 km/h. With some slick shifting of its standard five-speed manual gearbox (a four-speed slush box is on the way) it will run to 100 km/h from a standing start in a commendable 11.7 seconds, no mean feat for a weighty 1886 kg offroader.

Freeway cruising at over 100 km/h is a relatively effortless affair, the V8 ticking over with no undue sign of stress. But if you're prone to exercising your right foot, the Discovery will happily oblige and its occupants enjoy the aural rewards that only come from a hard-worked V8.

The gear change is clearly defined with good clear gates and a short throw, although the detent protecting reverse is easily beaten when selecting first. Once up to speed the shift is suprisingly light and snappy for an offroader, though once again it's caution all the way with first.

Offroad, the differentials and transfer box can be engaged while on the move, giving four "crawler" gears that come into their own on steep and slippery ascents. On trecherous descents low-range first allows the Discovery to slowly creep along at less than 10 km/h with all four wheels providing a formidable amount of engine braking without troublesome transmission snatch.

The high driving position gives the driver a clear panoramic view that helps spot overtaking opportunities when on-road and is great for peering over crests in critical offroad situations. As a whole it inspires a safe, efficient driving style and makes long haul journeys a much more relaxing affair than a conventional sedan car.

Only when pressing on does the Discovery's Range Rover-like chassis composure begin to limit forward progress. Employing such a long travel suspension it requires slight corrections of the steering wheel to keep it pointed down the straight and narrow. A steering damper, fitted to soak up steering wheel shocks in rough offroad conditions, hampers on-road feel. When punted hard the steering lacks precision around the straight-ahead and is easily deflected by broken bitumen. On smooth roads, however, the steering feels perfectly weighted and cannot be faulted.

Taken to the roughest bush tracks the Discovery displays outstanding ability and only when it's driven over roads that would bring any normal road car to its knees does the driver begin to appreciate its all round prowess. The long travel suspension dutifully soaks up irregularities and the body remains composed without any of the shakes or rattles that typified its predecessor, the archaic 110.

Discovery's handling also displays typical Range Rover parentage. In everyday driving it feels as good as its upmarket stablemate. Push hard and the Discovery begins to feel a little unsettled, with considerable body roll and consistent, but far from overwhelming, understeer.

Offroad, Discovery's handling comes into its own. There is a considerable amount of steering feel thanks to the aforementioned steering damper and the four-wheel drive chassis can be made to slew into tight corners with mild oversteer. It's always safe and entirely predictable but the Discovery's tendency to lean heavily mid-corner is potentially unnerving for passengers.

Because the Discovery is such an effective offroader it could, perhaps, be allowed some quirks as a car for transporting the whole family. While Europe gets both four-door and two-door versions of Land Rover's best ever we only get the two-door to play with, though with the dropping of the super luxury sales tax this may change. But trying to fault the Discovery's ability to carry anything up to seven in relative comfort isn't easy. Accommodation is excellent. Up front the seats are good but tend to lack under thigh support. The sculpted rear seat is perfectly acceptable for long journeys, more so than the Range Rover's, and the optional flip-down rear seats in the luggage area are comfortable even for adults.

The interior styling is the work of acclaimed British design house, Conran Design. It's obviously modelled on some of the more successful Japanese set-ups and is far and away the most user-friendly interior Land Rover has ever produced. Fit and finish is good, although the pale colour scheme looks disappointingly cheap.

For more than 20 years the go-anywhere Range Rover has rightfully worn the crown as the world's best offroader and despite the best engineering efforts and almost unlimited resources of the Japanese still holds its head high. But for many die-hard Aussie offroaders the dream of owning one of Britain's best has been fraught by high import tariffs and a plunging exchange rate. At $44,995 the Discovery isn't cheap but in terms of all-round ability it's a bargain.

GREG KABLE

AT A GLANCE

ENGINE: Fuel-injected, 3.5-litre 90-degree V8 developing 115 kW and 260Nm of torque.

SUSPENSION: Front: live axle, radius arms, Panhard rod, coils prings, telescopic dampers. Rear: live axle, radius arms, upper "A" frame, coil springs, telescopic dampers.

FUEL ECONOMY: Average 17.2 l/100kms.

PRICE: $44,995

LIKES: Performance, chassis dynamics, price.

DISLIKES: Nervous steering, cheap looking cabin

DISCOVERY UNDER FIRE

The Discovery could once call itself the best off-roader in its class, but the new Mitsubishi Shogun, with its complex four-wheel drive system, has its sights aimed at taking over the Land Rover's position. Two five-doors battle it out. Photography by Stan Papior

LAND ROVER'S PRODUCTS HAVE, over the years, carved a reputation for indestructibility in the most arduous conditions anywhere in the world, and the Discovery has been no exception. Beneath the glassy body with its Conran-designed interior, the parts that matter are just as rugged as those on the utilitarian Defender. Justifiably, the Discovery has been hailed as the best off-roader in its class.

But the new Mitsubishi Shogun has thrown that title into dispute. Now with a complex Super Select four-wheel drive system and changes aimed at improving its competitiveness in a growing market of sophisticated off-roaders, the latest Shogun comes armed with the right weapons to blow the Discovery out of its commanding position. In theory, at least.

On paper the Shogun emerges as the technical leader. While both off-roaders share separate chassis construction and 4wd with a choice of high and low ratios, and a centre diff to permit 4wd operation on or off-road, that's about all they have in common. The Land Rover's drive to all four wheels is permanent: the Shogun's is selectable, and it also has the benefit of a centre viscous coupling to ensure that torque always reaches the axle with the greater grip. In addition, the Shogun has centre and rear diff locks; the Discovery makes do with just one, in the centre.

Some would argue that the Shogun's Super Select transmission is unnecessarily complex, but it's hard to argue against the advantages. In isolating the front diff and propshaft from the drive, with the capability for it to be reselected at any time, there are fuel-saving and power-gaining opportunities to be had. But the distinct shortfall in power and torque compared with the Discovery V8i means the big five-door Shogun needs that extra power.

The Discovery gets 3.5 litres of injected all-alloy V8 to push along its considerable bulk, while the Shogun has an iron-block three-litre V6, also injected but cat-equipped. It's soon clear which engine has the easier task. With 164bhp at 4750rpm and 212lb ft of torque at a mere 2600rpm, the venerable Rover V8 has the clear advantage over Mitsubishi's 147bhp at 5000rpm and 174lb ft at 4000rpm. Straightforward overtaking manoeuvres for the gutsy Discovery demand advance planning for the Shogun, despite lower gearing to make best use of its power.

But where the Land Rover steals a bonus for its engine, it loses it all on refinement. It's not the engine that's objectionable, it's the transmission, which whines annoyingly and develops a rumbling vibration from deep within its bowels. Add to this the Discovery's pathetic door and window sealing, which produces substantial wind roar at speed, and the message is spelt out loud and clear: refinement is not the Land Rover's strongest suit.

In contrast, the Shogun's refinement is improved vastly over its predecessor. There's remarkably little wind roar, a refined, slop-free transmission, and tyre noise is virtually absent. Only the pronounced growl from the engine lets it down.

The Shogun's friendly road manners, with its good stability and light, relatively direct, trustworthy steering see it edge ahead of the

Axle articulation on Shogun (left) is no match for the Discovery's (below), though the Mitsubishi's rear diff lock gives it advantages over certain types of terrain

New Shogun looks dated and slightly agricultural inside (below, left) when set against the Conran-designed Discovery interior, though it feels more solid

LAND ROVER DISCOVERY V8i
ENGINE
Capacity 3528cc 8 cylinders in 90deg vee
Bore 89.0mm **Stroke** 71.0mm
Fuel and ignition electronic ignition and fuel injection
Max power 164bhp (122kW) at 4750rpm
Max torque 212lb ft(287Nm) at 2600rpm
SUSPENSION
Front live axle, radius arms, Panhard rod, coil springs, telescopic dampers
Rear live axle, radius arms, A-frame, coil springs, telescopic dampers
PERFORMANCE
0-60mph 11.7secs, 0-80mph 21.8secs, 30-70mph 12.4secs, max speed 105mph
FUEL CONSUMPTION
Test mpg 15.3
PRICE
£20,895

MITSUBISHI SHOGUN V6
ENGINE
Capacity 2972cc 6 cylinders in 60deg vee
Bore 91.1mm **Stroke** 76.0mm
Fuel and ignition electronic ignition, multi-point fuel injection
Max power 147bhp (110kW) at 5000rpm
Max torque 174lb ft(236Nm) at 4000rpm
SUSPENSION
Front independent, double wishbones, torsion bars, anti-roll bar, coil springs, telescopic dampers
Rear live axle, radius arms, Panhard rod, coil springs, telescopic dampers, anti-roll bar
PERFORMANCE
0-60mph 13.1secs, 0-80mph 25.6secs, 30-70mph 14.2secs, max speed 104mph
FUEL CONSUMPTION
Test mpg 17.2
PRICE
£22,349

Discovery. It makes a refreshing change to the Discovery's characteristic sloppiness, which is accompanied by lurching and swaying round twists and turns. Likewise, the Shogun pulls up straight and true: the Discovery can get out of line under heavy braking and, unlike the Shogun, there's no anti-lock option, although its brakes wash off speed quickly.

In spite of its narrower tyres, the Discovery grips better, but ultimately pushes the nose wide. The Shogun is better balanced, but transfers from benign understeer to a mild rearward bias on the limit, should you ever want to drive these beasts in such a fashion.

On the road, the Shogun's better refinement and more positive road behaviour make it generally the more pleasant of the two. But although the ride is generally good, it's not quite as supple as that of the Land Rover, and the tendency for the long-wheelbase Shogun to shake, shudder and flex is a significant flaw.

Off-road, the Discovery has lost none of its supremacy. Compared with the Shogun it rides better; axle articulation is considerably greater; it has a beefier engine, which is able to pull it through deep mud or up steep inclines on a whiff of throttle; and engine braking is greater on descents. The gearing in low range is lower, too, enabling cautious progress through tricky sections, and the suspension is more progressive and has longer travel.

However, the Discovery's soft suspension means the body lurches more over severe bumps at extremely low speeds, and the lack of a rear diff lock can, in certain situations, lead to a loss of traction where the Shogun can power through with its rear and centre diffs locked. But the Shogun is limited by restricted axle articulation, especially at the front. It lifts a wheel where the Land Rover keeps all the wheels firmly on the ground.

Dynamically, the Discovery is still the car to beat off-road, but the Shogun's more positive road behaviour and fine refinement give it the edge on tarmac. It's also more economical. In 400 miles of road testing, including off-road, the Shogun returned 17.2mpg against the Land Rover's 15.3mpg.

Check the price lists and the Land Rover V8i five-door appears to offer the better value at £20,895, some £1454 less than the Shogun V6 five-door. But though the Discovery comes with the basic requirements of electric windows, central locking, power steering and alloy wheels, the Shogun offers this and more. As standard it also has cruise control, twin heaters, and a large, electrically operated sliding sunroof (not available on the Discovery).

Each of these gargantuan off-roaders seats seven, so space isn't a problem. The Discovery has the better-designed cabin, though, with acres of rubber mat-covered oddment space, and the rear-most seats fold away unobtrusively, taking up hardly any load space. The centre row of seats splits and folds, too. The Shogun's bulky rear seats might look an after-thought when folded up, but its two rear rows of seats fold flat to create a sizeable double bed.

A fresh outlook to interior design characterises the Land Rover. It is bright and airy inside, but there's plastic everywhere, and it sometimes feels flimsy. Even so, this is an improvement on the Shogun cabin's dated, even slightly agricultural look. The angular, shiny plastic facia, liberally splattered with switchgear, is particularly offensive, but it feels the more solid of the two.

Overall, the Discovery still takes the laurels off-road but, for road driving, the Shogun is the preferred choice for its greater refinement and ease of driving. Not that the Shogun gets it all its own way. After all, the Discovery has the better engine and the better ride. It still has immense appeal, and rightly so, but these days it's harder than ever to make the choice when the Shogun is such a capable and attractive alternative. ■

DISCOVERY TEST

Gallic trial

By Tim Webster

A FIVE door Discovery was always 'on the cards' right from the start, and in typical fashion, Land Rover broadened Discovery's appeal in a single stroke. Tim Webster drove across France putting the latest model to the test.

YOU may have gathered from last month's report on the French Cevennol Trophy that part of the magic of the event was provided by our chosen steed — the latest five door Discovery.

All in all, we couldn't have chosen a better vehicle as a 'reporter's chase car', and also as a high speed Autoroute express. The Discovery excelled in both guises, delivering us there and back safely, and with remarkable economy.

So much so that my companion, not normally given to tramping car showrooms, has actally taken to gazing longingly at dealers' Discoverys in the hope that one might replace our motley Land Rover/Range Rover fleet.

I suspect that many other prospective buyers may find themselves in the same situation. Especially as the five door Discovery, lacking the body decals of the three door model, but demonstrating a more feline bodyline, has equal appeal to both sexes. Indeed, during our time with the test Discovery, it was equally referred to as 'handsome' and 'cute' — factors that no doubt won't have escaped the more switched-on dealers.

But this latest Discovery is likely to have to undergo the time honoured 'test by familiarisation' that so characterises the launch of any new Land Rover model. Defenders, Range Rovers and Discoverys are multi-facetted vehicles — a virtue of the dual role so many play. Our test Discovery could cruise at an indicated 95mph all day in relative silence. And then, with the flick of its transfer box lever, climb the side of a mountain. How can you find out all there is to know about a vehicle like that in a year, let alone just a week?

We knew it would be a problem for us from the outset. Climbing into the Discovery, and taking in the unfamiliar new interior colourway — aptly titled Bahama Beige — we settled down to a view over the machine's well contoured bonnet, and headed for France. With a reputedly tough French rally to 'chase', 1000 kms of varied French countryside to drive, and the appeal of three days basking in Provencal sunshine, it would be easy to disregard the Discovery in favour of the ever changing Gallic landscape. Or would it?

The first surprise was the muted strength of the 200 Tdi direct injection diesel engine. Acceleration through the gears was brisk, initial changes being made as the governor came into operation rather than scrutinising the dash mounted rev. counter for the best change-point.

That soon changed in favour of token economy for a while, a move that paid dividends as fuel stops on both sides of the Channel came into view . . . and were ignored. Indeed, we recorded around 35mpg overall — far better than we would have expected, and a result that then prompted flat out all day cruising to reach Provence. We suspected the fuel economy would plummet from there on; it didn't — we still recorded 35mpg despite my leaden foot!

We also realised that, for the moment, a five door Discovery makes it impossible to travel incognito. Stopping in the UK for quick breathers and necessary replenishment of the mandatory Coca-Colas and Snickers bars brought on the unwanted attentions of onlookers — Land Rover enthusiasts who must lurk, habitually, behind the fuel pumps at Corley, Watford Gap and Toddington motorway services, ready to pounce.

We fought off most enquiries. 'Has it got air-conditioning?' (yes, it has, even on the Tdi version); 'What's it like with five doors?' (we couldn't answer that one!), and even 'Can I sit in the back for a moment?' We found at that point, much to our surprise, that you can get a chirp of wheelspin and quite remarkable standing start acceleration out of the 200 Tdi engine — something we were to prove time and time again as we headed southwards.

In France, the Discovery probably drew more attention than it did in England. After all, it was, at times, the fastest vehicle on the autoroute, despite carrying the two of us and the sum total of my off-roading recovery kit that I needed to fit heavy duty suspension to my Ninety to carry.

And, as the ambient temperatures of France rose, so the air-conditioning came more into its own, occasionally supplemented by one or both of the twin sunroofs. Being pretty undecided as to whether I like air-conditioning or not, I did notice that, during the long Peage hours, it readily sent my co-driver off into her dreams. And kept me awake hour after hour with a clear head, so I had time and freedom to glance around the Discovery's interior.

In five door form, the trim is well balanced (or perhaps co-ordinated would be a better world), and complements the status and price of the Discovery well. Look aft, and you're reminded just how big the Discovery is. But the light beige of the interior also helps to give it an airy feel and, to my delight, shrugs off the inevitable soiling of off-roading excursions.

I also noted and appreciated the clarity of the instrument panel, the balanced sensitivity of the power steering, and the support of the *very* comfortable seats.

I also commented freely (or so I'm told) on how useful the centrally mounted switchgear was for the electric windows, how quickly they wound down and up . . . and how powerful they were by virtue of my tendency to drive one-handedly, gripping the rain gutter with my free hand. Until, that is, someone carelessly decides I'm creating too much draft and tries to shut the window without telling me.

As a road car, there is no doubt that Discovery delights in working hard for you. In Europe, there's absolutely no disgrace in running in diesel, and the 200 Tdi is peppy enough to keep up with the French traffic flow, yet help you escape the inevitable traffic prangs that a foray into Aix-en-Provence brings daily.

Parking is another matter, since a small crowd collects around the car in minutes, drawing indecipherable comment all ending with that characteristic Provencal '-ang', that Peter Mayle so accurately describes in his books. It's fun to sit there in the car being admired; it's not so much fun to inadvertently pull away when the French are inspecting your rear axle.

But if everyday France was as tough a road trial as we could have found for the Discovery, worse was yet to come. Chasing the Cevennol Trophy meant running the same tracks that the competitors used. All this with a man behind the wheel who habitually uses nature's best trees to inscribe his Ninety with map-relief panelwork.

Of course, we were on best behaviour, and Discovery responded to a light touch and sensibly driven tracks with a total clear. Some competitors weren't so lucky, and managed to sandwich Suzukis, roll Toyotas and drown G-Wagens in quick succession.

Even on standard OE Goodyear Wranglers, the Discovery is an off-road professional. Hump the LT77 gearbox into 'third low', leave the engine at tickover, and the machine will haul itself up almost every slope. Where Jeep Cherokees waved wheels, Discovery will notify you that no more axle travel is available with a gentle thump onto its bump stops. But *its* tyres won't leave the ground.

There are times where you have to be circumspect. Discovery's barrel-sides *are* vulnerable to scratches and Natures's side swipes from trees. The solid Eskdale Green paint finish of our machine bore its punishment well, and responded first time to a courteous wipe with T-Cut; our French colleagues had to be content with branches taking away entire panels on the same routes.

But to understand what Discovery is all about, you have to push it to the edges of its 'performance envelope'. It's with some reluctance that I'll claim we did, on and off-road all that week in France. And the Discovery bounced back for more.

Now, some weeks after we returned from Provence and the Cevennes, I'm still convinced that I know what makes Discovery so good. It's because it really *is* a dual purpose machine. Its good looks relaxed me at times when it was important to make the right impression. Yet, the oval Land Rover badge in the centre of the steering wheel prompted a spirited try at some of the toughest off-road tracks I've yet to encounter.

Put simply, we liked it. We liked it a lot!

DISCOVERY TEST

81

ROAD TEST

DISCOVERY V8i 5-DR

With European pricing, its faults would be insignificant: it's a rattling good buy. Even with 120 per cent import tax, it'll still sell...

BY THE time we hand back our test cars, at the end of a two-week driving stint and a formal performance assessment, we've usually reached some firm conclusions on their pros and cons, and also on their competitive position in the market. But with a few, we have mixed feelings and the Land-Rover Discovery V8i is one of them.

Although it has defects, they're not serious and we have no qualms in naming it the best leisure-type off-road vehicle we've yet tested. Which concurs with most reactions in Europe, where sales have outstripped predictions and now account for one third of all 4x4 production at the Solihull plant.

Our confusion arises from the price, for while in South Africa the Discovery ranks as an expensive luxury – its

KEY FIGURES	
Maximum speed	169 km/h
1 km sprint	35,5 sec
Fuel tank capacity	81 litres
**Litres/100 km at 100	10,53
Optimum fuel range at 100	769 km
*Fuel index	14,74
Engine revs per km	1 482
Odometer error	0,5 per cent under
National list price	R179 607
(*Consumption at 100 plus 40%)	
(**Factory claimed figure)	

R179 607 launch price inflated by heavy import taxes – in Europe, it's an affordable, leisure-lifestyle vehicle that undercuts and out-performs the Japanese 4x4s it was built to challenge. Even so, those South Africans who can afford to buy it are likely to regard it as comparatively sound value, listed R63 000 below the cheapest Range Rover (which is, of course, also imported).

In Britain, this five-door, V8-powered Discovery recently retailed for 21 000 pounds and oddly enough, the alternative TDi model, propelled by an intercooled, direct injection turbo diesel, is the bargain buy over there, tagged 800 pounds cheaper. In South Africa, it's the other way round, for the TDi costs nearly R5 000 more. So our local perspective is distorted, price-wise, and when the vehicle's weaker points are considered, as will occur in this report, they should be seen in relation to the pricing in Europe, where the Discovery is seen as a rattling good buy.

It was developed in direct reaction to market developments. Although Range Rover dominates the luxury sector and the venerable Landies still rank at or near the top of the "genuine workhorse" market, the company had no answer to the invasion of Japanese shock troops – the attractively styled, easy-to-drive leisure off-roaders that were cheap enough to be bought as weekend/second cars by middleweight earners. As these sales, dominated by the Mitsubishi 3,0 Shogun (Pajero) and the Isuzu Trooper and with competition from Toyota and Nissan, were growing at a booming pace, Land-Rover badly needed to participate, to secure its future prospects.

The Discovery was the result – a stylish and handily sized, short wheelbase wagon which features luxury specs, is backed by a comprehensive accessory programme and was priced at only 15 750 pounds, in three-door form (with either powertrain), at its 1989 launch. Unlike the Range Rover, where buyers had to wait almost 15 years to buy a model with two additional doors, the five-door followed just 12 months later. The problems faced by the already outclassed Japanese contenders were compounded by the fact that the 3,5-litre all-alloy V8 was now fuel-injected, enhancing economy as well as performance.

700 UNITS A WEEK

It is this later version, the V8i, and its five-door, turbo-diesel stablemate, the TDi, that have come to South Africa,

Distinctively styled with attractive alloy wheels, the Discovery combines fast, comfortable cruising capabilities with very easy and refined off-road travel. It is powered by a 3,5-litre fuel-injected, alloy V8 (right) and has an airy, modish interior (far right) accommodating five or seven people. The driving position is good, providing a commanding view of the road, but the air-con/heater controls are confusing and awkward to reach.

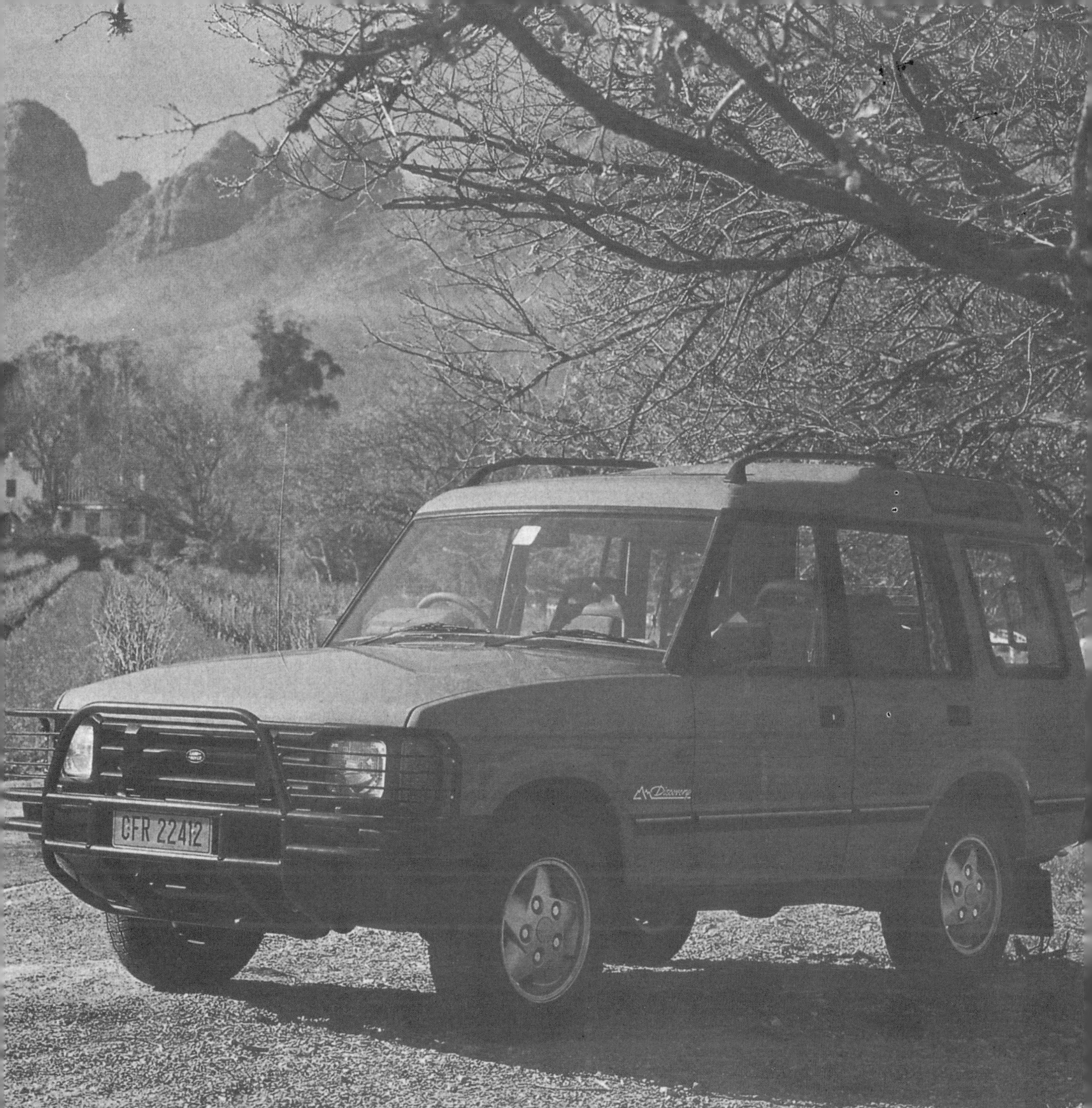
CFR 22412
Discovery

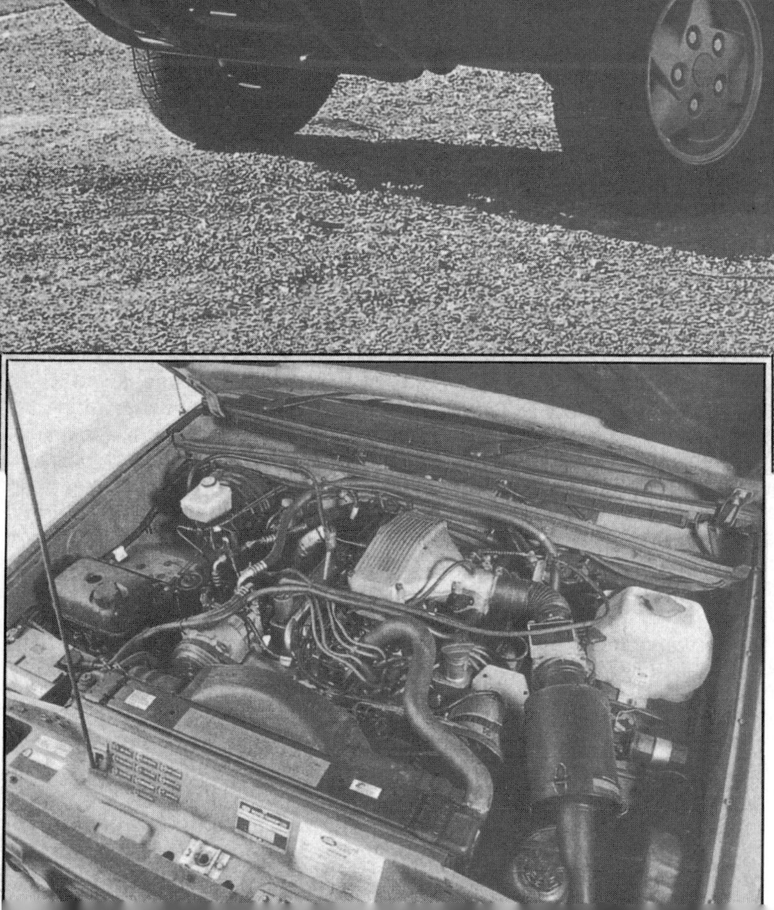

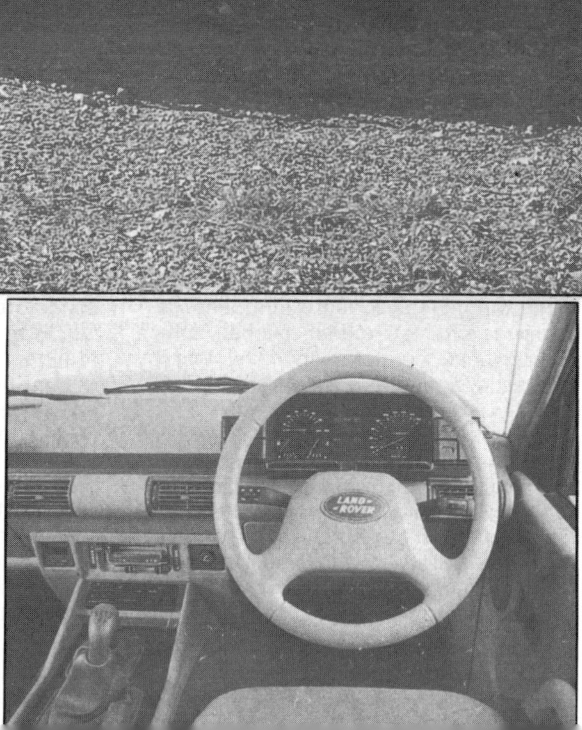
LAND ROVER

The test vehicle was upholstered in tweedy cloth and the seats were well shaped and generous in size. The longer lever controls a five-speed gearbox and the stubby one in front of it, the two-speed transfer 'box and the diff lock, which can be engaged on the move.

There is 744 dm³ of station-wagon type luggage space, increasing to 1 448 dm³ with the 60/40 back seat fully folded.

and this is the first market outside Europe and the Middle East to get them. Rover's Solihull factory is now building 700 Discoveries a week, more than double the initial rate, and one of its directors, Chris Woodwark, recently crowed that "this is the only British vehicle to win back, so decisively, a market that had once been lost to Japanese competition."

As we inspected our plain, rather dull grey test vehicle after delivery, we realised it was actually larger than it had first appeared, or felt, during our launch test drives. In fact, its 1 979 kg licensing mass makes it 164 kg heavier than the Range Rover we tested (in 3,5-litre form) in 1984, as well as 51 mm longer, 7 mm wider and 148 mm higher. It is also 80 kg heavier than the Land-Rover 110 wagon we tested in March last year (though that had a longer wheelbase) and more than 100 mm longer than the Mercedes G-Wagen we assessed for our May 1989 issue (although that had a much longer wheelbase and was 21 kg heavier).

On or off-road, the Discovery feels more compact and wieldy than a Range Rover; and although the dearer vehicle is likely to remain the top image 4x4, on snob appeal as much as anything else, the newcomer is bound to win many converts. It was designed after a thorough analysis of the European/British leisure off-roader market, which had been created and developed by the Japanese and which Land-Rover knew little about.

Rover briefed a survey team to research it and determine and prioritise customer needs, and the designers were apparently given only one restraint: the new vehicle must be neither an under-spec'd Range Rover nor an up-spec'd Landie. In the event, the Discovery has been built on engineering strengths established in four decades of L-R marketing, for it uses an alloy-panelled body mounted on a chassis which is essentially the same as the Range Rover's. It has exactly the same wheelbase, long-travel, coil-sprung, live axle suspension and permanent four-wheel drive.

The five manually-selected forward gears drive through a two-speed, centrally located transfer box and when the going gets really sticky, the differentials can be locked and/or the "crawler" ratios engaged on the move, making the vehicle virtually unstoppable over any terrain feasible for 4x4 travel. With the normal high ratio selected, on the other hand, the gearing is long and the fuel-injected V8, impressively muscular and smooth in its lower ranges, provides quiet and effortless open-road touring.

As used in the current Range Rover Vogues (not yet tested by CAR) the alloy V8 has been uprated to 3 947 cm³ and develops 134 kW at 4 750 r/min. For the Discovery V8i, the motor retains the classic 3 528 cm³ capacity but uses a sophisticated electronic management system co-ordinating fuel injection and ignition systems.

MORE POWER, TORQUE

In this latest form it produces 122 kW at 4 750 – 13 per cent more power than the carburettor-fed 3,5-litre – and peak torque is up by 10 per cent to 287 N.m at 2 600 r/min. Using a 0,77:1 fifth gear and shod with 205/16 Pirellis on 16-inch stylish alloy wheels, it has an overall gearing (in high ratio) of 40,57 km/h per thousand revs, enabling it to cruise at the 120 km/h limit with just 3 000 rung up on the tachometer. tachometer.

The brakes are 299 mm discs, ventilated in front and solid at the rear, and the steering is by a power-assisted, recirculating ball system controlled by a fat-rim, stylish, twin-spoke wheel that blends with the airy, "modern lifestyle" interior but, as we shall see, is designed with the accent on appearance rather than practicality. Initially available only in pale blue but now with a beige alternative, the chic interior, created by Conran Design to "out-mod" the Japanese, has been criticised for lacking the overt class of the Range Rover. But it clearly suits the target market and, of course, the (European) pricing.

It is certainly spacious with exceptional head and leg room and the four doors permit easy access from either side. The rear door swings wide, too, on its right-hand hinges, but when the pvc-clad spare wheel is bolted vertically in place on it, it significantly obscures the driver's direct rear view.

He (or she) then becomes more reliant on the large, electrically-aimed door mirrors which are included in the standard spec, along with central locking, air-conditioning, power windows, rear screen heat, wash 'n wipe, rubber mats for console and facia tops and Alpine roof panels – which, in conjunction with a raised roof line, account for much of the airy effect achieved in the interior. AAD list a wide range of accessories with strong appeal to leisure off-roaders and the test car was equipped with twin tip-or-remove screened glass "sun hatches", inward-facing "jump seats" on either side of the rear luggage space (which will take adults for reasonably long trips and fold neatly away when not in use), a robust brush guard and a roller-blind stowage space cover.

There was also a neat, easily portable radio/tape which worked well but whose vertical mount at the base of the centre facia made the tape slot vulnerable to dust intrusion. Not available for inspection were the waterproof

The two inward-facing "jump seats" are optional extras and fold out of the way when not in use. The rear door swings wide on vertical hinges (right).

seat covers offered to protect the cloth upholstery from damage in serious off-road excursions, or the rear-mounted clamp-on roof rack which had seemed prone to rattles, on one of the vehicles used at the Paarl launch.

However, the front roof section is fitted with fixed rails and with the addition of cross runners stored inside the vehicle, there is enough capacity for bulky loads, in addition to the wagon-type space available inside, to satisfy most owners. Assessed using our metric block system, the Discovery's luggage space came out at 744 dm³ with 60/40 split rear bench seat erected, but with it folded flat and tumbled forward, which is easy to do (although one catch had disassembled itself, on the test car), the capacity increased to 1 488 dm³.

Interior stowage space is another strong point in the interior, with rubber-lined shelf atop the facia, map pockets above the sun visors, large pockets in front and rear doors, others on the backs of the front seats and ceiling nets to hold maps or documents. The front bucket seats (which have rotating rake adjust) and rear bench seats are well shaped and generously sized and the high mounted body provides driver and passengers with a commanding view of the road and surrounding countryside.

Although most people like the colour co-ordinated interior, a few find it too modish for their taste. The tweedy blue cloth upholstery in the test vehicle was used with shiny plastic trim that scratched easily and pale blue carpeting.

The ergonomics are generally good, with instruments clearly displayed in a facia-top binnacle in front of the driver and all controls within easy reach, except for the centrally placed heat and ventilation panel. This typifies the style-above-practicality approach that weakens some aspects of the interior design. For while the array of controls look crisply bright and attractive, they are confusing to use until you have studied the handbook. Once we had done that, we extracted a strong and versatile performance to heat, ventilate or cool; but it remained irritating to have to place the short, vertically-moving sliders in the centre position, in order to obtain a good airflow.

The stylish twin-spoke steering wheel, too, has its grip dimples at top and bottom instead of in the natural driving positions and other poor interior features are the fiddly main lighting switch built into the left-hand stalk control and the stalk-tip mounted hooter button, a feature we hoped we had seen the last of, in currently built cars. In contrast, the right-hand stalk provides a neat variable speed intermittent wipe control that is of real benefit, in a vehicle of this type.

The real charm of the Discovery lies in its all-rounder capabilities, for it drives and travels like a large and comfortable car, while making off-road travel exceptionally easy and comfortable. Though its overall dimensions and weight outrank the Range Rover, it *seems* smaller and lighter and many owners will happily use it seven days a week, for all-purpose transport.

LAID-BACK V8

The V8 comes to life instantly, once you've managed to insert the ignition key in its badly obscured recess, and burbles away in laid-back fashion in every situation from freeway cruising to walking pace propulsion across really rough terrain. Husky and smooth in highway usage, it produces a seemingly limitless supply of low speed torque, controlled by a remarkably progressive, long-travel throttle. Tractability is so good, it will pull from 15 km/h in fifth gear without snatch, then accelerate smoothly without protest.

As it's harnessed to long gearing which it pulls with consummate ease, there's little point in exceeding 4 000 r/min, which is just as well, since it quickly runs out of steam above that point and sounds less happy, as the revs rise. Though not profiled for performance, the powertrain will hustle the vehicle to a top speed of 169 km/h on a flat road – 13 faster than the 3,5-litre Range Rover we tested five years ago – and will accelerate from standstill to 100 in 14,4 seconds – much faster than any of the three 4x4s quoted in our earlier size and weight comparisons.

As we were unable to use our electronic test gear due to technical problems, we resorted to stop watches paired with carefully calibrated instrumentation, to get our performance results; and the fuel figures quoted in our tables are "factory claimed" specs, obtained in the course of an EEC type programme.

In our two-week driving spell we logged an overall consumption of 15,5 ℓ/100 km, rather heavier than the fuel index figure of 14,74, but this can be explained by the inclusion of the performance tests and by hard driving, which most Discovery owners are unlikely to emulate.

We found the vehicle would sprint to the 1 km post in 35,5 seconds, reaching a terminal speed of 160 km/h, but the most impressive results came in our overtaking acceleration tests. The bulky, 1 979 kg vehicle would surge from 80 to 100 in third gear in 4,5 seconds, for example, or from 100 to 120 in fourth in 6,9. It is these figures, and the others in our "overtaking" table, that log the Discovery's strongest performance asset – its excellent throttle response that makes it fun to drive on the road.

As you can boot past slower moving vehicles or up gradients with relative ease, open road cruising is much more practicable than in most other 4x4s, accompanied by the clipping rhythm of the injected V8 and with a quiet and comfortable ride reminiscent of the Range Rover. Although the long travel coils are restrained by effective damping, the ride can feel floaty over some surfaces and the vehicle leans heavily through fast corners, which needs getting used to. It also tends to rock very slightly laterally on some roads. But all in all, the suspension offers the best available compromise to accommodate serious off-road usage, when the Discovery will ripple over rough stuff that would jolt, strain or stop some current 4x4s.

AMBLING COMPETENCE

The short wheelbase, high ground clearance and exceptional entry/egress angles built into the chassis enable the Discovery to amble through trenches and across steeply defined gradients that would cause most 4x4 bakkies to struggle. The off-road handling is in fact terrific, with enough steering feel to give complete confidence.

In heavy going, the vehicle can be manipulated by the combination of precise throttle control, steering input and brakes to place it exactly where you want it. Although the handbook suggests engaging the diff lock (which is operated by the same stubby shift used for the transfer 'box) for wet grass, snow, mud or sand, it's rarely required in practice and in any case, can easily be engaged on the move.

Blunt and functional, the Discovery carries its spare wheel vertically on the rear door where it partly obscures the driver's view. It is wider, taller and longer than the Range Rover.

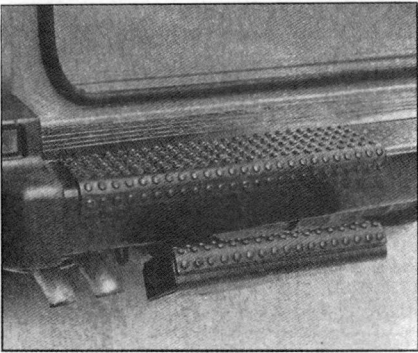

This rugged steel step (above) is spring loaded to lift it out of the way when off-road.

It simplifies rear entry (or loading operations) (right) when the jump seats are in use.

Warning lights are provided for transfer 'box lubricant temperature and diff lock operation and if the latter one remains on after usage, the transmission must be "unwound" quite simply by reversing for a short distance. With diffs locked, the Discovery will walk down a steep gradient quite steadily with all wheels transmitting engine braking and no transmission snatch.

Although reverse gear can be difficult to find, the stiffish, short-throw gearshift has a positive action and improves on the move, producing surprisingly snappy shifts for a 4x4. The clutch is heavier than some would like but is clean-acting and progressive.

The power steering, which feels rather vague in the straight-ahead position and is prone to shimmy over rutted tarmac, comes into its own off-road, when the damper seems to work harder and the body retains its composure even when the road wheels are travelling vigorously over rough terrain.

On road, the long travel coils and solid front axle exact their penalty in brisk, hard-driven travelling, marked by pronounced understeer and the considerable body roll. If you negotiate a tight bend too rapidly you'll feel both inside wheels lose traction until backing-off the throttle tightens the line, without drama, but on a smooth or damp road, you can push the back end gently outwards, under power, as the V8 growls and the body leans. Though a reasonably experienced driver will know such behaviour is completely safe, his passengers are apt to reach for the strategically-placed grab handles.

The powerful brakes are at their best off-road, when they give effortless control with moderate pedal pressure, but in our ten-stop emergency programme they were apt to lock-up in random fashion, forcing us to back-off and affecting both stopping times and directional stability. As our graph shows, the behaviour was erratic and the resultant average time taken to stop from 100 km/h was 4,21 seconds; inside the safety ball park, for a vehicle of this type, but certainly not impressive.

TEST SUMMARY

Although it carries a heavy price penalty in this market, the Discovery's remarkable combination of on-and-off-road competence well merits its position near the top of the CAR Guide 4x4 section, which is listed by price. Refined, spacious and with luxurious ride comfort, it can be happily used for about-town business, motorway cruising or in exacting wilderness conditions.

Although few owners may explore its off-road capabilities, those that do will find them reassuringly easy to use. Those that don't can enjoy its leisure-lifestyle image, confident in the knowledge that their vehicle *does* have what it takes, should it ever be needed.

SPECIFICATIONS

ENGINE:
- Cylinders V8
- Fuel supply ...... electronic fuel injection
- Bore/stroke................... 88,9/71,1 mm
- Cubic capacity .... 3 528 cm³
- Compression ratio .... 9,35 to 1
- Valve gear .... o-h-v
- Ignition .... electronic management
- Main bearings .... five
- Fuel requirement .... 97-octane Coast
  93-octane Reef
- Cooling.................................... water

ENGINE OUTPUT:
- Max power DIN (kW).................... 122
- Power peak (r/min).................... 4 750
- Max usable r/min 5 500
- Max torque (N.m) 287
- Torque peak (r/min) 2 600

TRANSMISSION:
- Forward speeds...................... five plus transfer ratio
- Gearshift................................ console
- Low gear............................. 3,321 to 1
- 2nd gear .... 2,132 to 1
- 3rd gear .... 1,397 to 1
- 4th gear .... 1,000 to 1
- Top gear .... 0,770 to 1
- Reverse gear .... 3,429 to 1
- Final drive .... 3,538 to 1
- Drive wheels .... front and rear
- Transfer ratios .... 1,222:1 (high)
  (lockable centre differential) 3,320:1 (low)

WHEELS AND TYRES:
- Road wheels.................. alloy (option)
- Rim width .... 7 J
- Tyre make .... Pirelli Akros Scorpion
- Tyre size............................... 205 SR 16
- Tyre pressures (front)............. 190 kPa
- Tyre pressures (rear) .... 260 kPa

BRAKES:
- Front.............. 299 mm ventilated discs
- Rear............................... 299 mm discs
- Hydraulics .... split circuit
- Boosting................................. vacuum
- Handbrake position........ between seats

STEERING:
- Type .... power-assisted recirculating ball
- Lock to lock .... 3,8 turns
- Turning circle..................... 11,9 metres

MEASUREMENTS:
- Length overall...................... 4 521 mm
- Width overall....................... 1 793 mm
- Height overall .... 1 928 mm
- Wheelbase......................... 2 540 mm
- Front track .... 1 486 mm
- Rear track........................... 1 486 mm
- Ground clearance................... 253 mm
- Licensing mass .... 1 979 kg
- Mass as tested .... 1 980 kg

SUSPENSION:
- Front................................. live axle
- Type .... coils, radius arms, Panhard rod
- Rear................................. live axle
- Type .... coils, radius arms, upper A frame

CAPACITIES:
- Seating... 5/7
- Fuel tank .... 81 litres
- Luggage trunk .... 744 dm³
- Utility space .... 1 448 dm³

WARRANTY:
12 months, unlimited km

TEST CAR FROM:
Associated Automotive Distributors (AAD)

TEST RESULTS — DISCOVERY V8i 5-DR

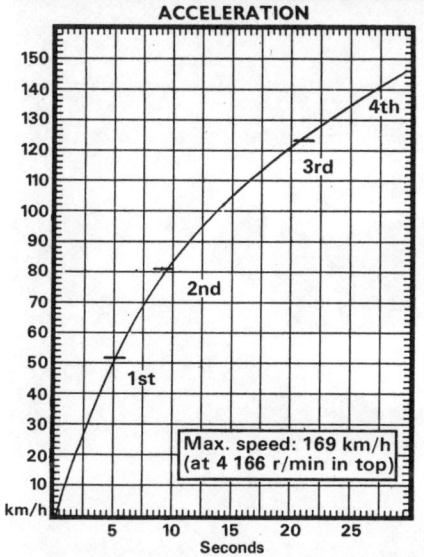

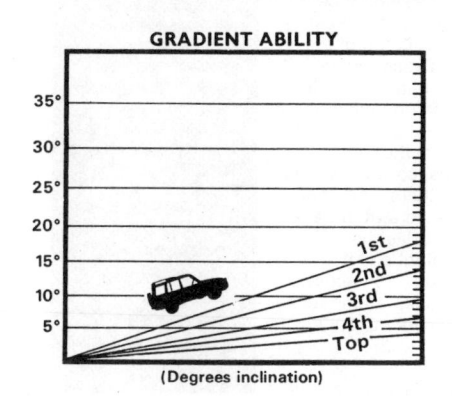

ACCELERATION

Max. speed: 169 km/h (at 4 166 r/min in top)

PERFORMANCE FACTORS:
- Power/mass (W/kg) net 61,64
- Frontal area (m²) 3,45
- km/h per 1 000 r/min (top) ... 40,57

(Calculated on licensing mass, gross frontal area, gearing and I.S.O. power output.)

TEST CONDITIONS:
- Altitude .... at sea level
- Weather .... warm, windless
- Fuel used .... 97 octane
- Test car's odometer 6 888

GRADIENT ABILITY
(Degrees inclination)

MAXIMUM SPEED (km/h):
- True speed 169
- Speedometer reading 178

(Average of runs both ways on a level road.)

Calibration:
Indicated:	60	80	100	120
True speed:	56	75	93	110

ACCELERATION (seconds):
- 0-60 6,4
- 0-80 9,0
- 0-100 14,4
- 0-120 20,0
- 1 km sprint 35,5
- Terminal speed .... 160,0 km/h

OVERTAKING ACCELERATION:
	3rd	4th	Top
40-60	3,8	4,5	9,1
60-80	3,5	4,8	8,3
80-100	4,5	6,4	21,0
100-120	6,2	6,9	18,6

FUEL CONSUMPTION (litres/100 km):
- Urban 19,08*
- 100 km/h 10,53*
- 120 km/h 14,92*

(*Factory claimed figures)

BRAKING TEST:
From 100 km/h
- Best stop 3,5
- Worst stop 5,0
- Average 4,21

(Measured in seconds with stops from true speeds at 30-second intervals on a good bitumenised surface.)

GRADIENTS IN GEARS:
(high ratio)
- Low gear .... 1 in 3,1
- 2nd gear .... 1 in 4,0
- 3rd gear .... 1 in 5,9
- 4th gear .... 1 in 8,3
- Top gear .... 1 in 12,5

(Tabulated from Tapley (x gravity) readings, car carrying test crew of two and standard test equipment.)

GEARED SPEEDS (km/h):
Low gear	45*	52
2nd gear	70*	81
3rd gear	106*	123
4th gear	148*	172
Top gear	193*	223

(Calculated at engine power peak* – 4 750 r/min and at max. usable r/min – 5 500 r/min.)

INTERIOR NOISE LEVELS:
	Mech.	Wind	Road
Idling	48	–	–
60	62	–	–
80	63	72	65
100	68	76	67

(Measured in decibels, "A" weighting, averaging runs both ways on a level road: "mechanical" with car closed; "wind" with one window fully open; "road" on a coarse road surface.)

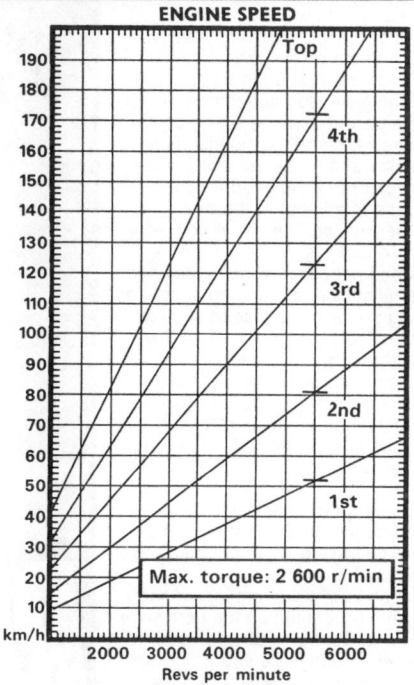

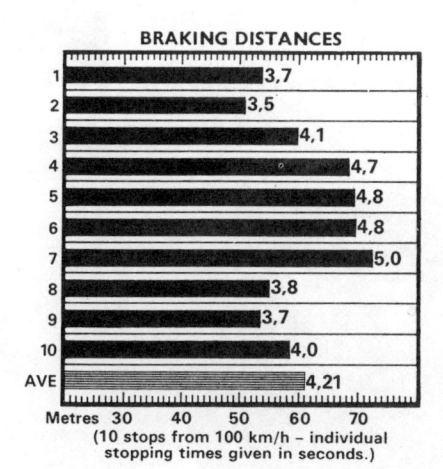

ENGINE SPEED

Max. torque: 2 600 r/min

BRAKING DISTANCES
#	Seconds
1	3,7
2	3,5
3	4,1
4	4,7
5	4,8
6	4,8
7	5,0
8	3,8
9	3,7
10	4,0
AVE	4,21

(10 stops from 100 km/h – individual stopping times given in seconds.)

MUDDY, OR NOT MUDDY? ASK this question of four-wheel drivers and too often the answer is that a dab of dirt will be just fine, so long as people can still read the white lettering on the tyres.

Many off-roaders are bought as fancy dress to make urban driveways seem more windswept and interesting, and to give owners that bush-hat-and-machete feeling as they scoot between Tesco and the disco. Few stray very far from Mr Macadam's tar-sealed surfaces, and when they do, it's normally to territory which wouldn't trouble a front-drive hatchback.

Despite the fact that most owners stick to the firmest of terra, manufacturers continue to make their off-road vehicles ever more manageable in the mire. Supercar makers face a similar irony, developing cars of such immense capability that just a handful of owners will have the skill or yet the opportunity to use their potential. But 'I could if I wanted to,' is a powerful buying incentive; sales of 4x4s and supercars are holding up well in these straitened times. And what other reason could there be for running around in something as expensive, complex, slothful, heavy, unwieldy and thirsty as a 4x4, when you don't have the need for one?

Come to think of it, there is another reason. It's fun. A complete hoot, in fact.

I don't necessarily mean leaping from hillocks and plunging at speed through door-handle-deep scrunge, as practised by the All Wheel Drive Club (though I can see the enormous attraction of that, too). There's also the gentler approach, the challenge of tackling remote, arduous terrain, the thrill of making the car do seemingly impossible things. The satisfaction of thinking your way around, and overcoming, an obstacle. And then there's the plain, childish pleasure of messing about in the mud.

Which is why we're bashing north on the A1, destination the Perthshire Off-Road Centre - 4000 acres of Scottish countryside and a purpose-built training course, a few miles outside Crieff. Ostensibly, we're comparing the virtues and vices of a trio of long-wheelbase, turbodiesel off-roaders. But secretly, we're off to play. Our toys are the Land Rover

RUNNING AMUCK

Brett Fraser, in the mood for yomping, takes a Vauxhall Frontera, Land Rover Discovery and Mitsubishi Shogun to Perthshire   PHOTOGRAPHS BY TIM ANDREW

Discovery TDi, Mitsubishi Shogun turbodiesel, and, newest of them, the Vauxhall Frontera turbodiesel.

Although it's the freshest face in the field, the Frontera is the least mechanically sophisticated of the group. Leaf springs suspend its rigid rear axle and it has no centre differential, so it can't be used on the road in four-wheel drive. It does, however, have automatic free-wheeling front hubs, so when you're done in the rough, there's no need to get out of the car to switch back to two-wheel (rear) drive. Its simplicity - some might call it crudity - is deliberate; it's meant as a sub-Discovery model, and it sells for a very sub-Discovery £16,830; the Solihull 4x4 is £21,025.

At the other end of the mechanical evolutionary scale is the

Three cars ride over river bed (above) travelling in price order: cheapest, the Frontera, ahead of Discovery and Shogun. Discovery, bogged in peat, about to be pumped up on exhaust-blown inflatable bag (left), under eyes of (L to R) Helena and Sandy Bell of Perthshire Off-Road Centre, and author Fraser

Shogun (top) limited off-road by lack of power, and by excess rear overhang so back end fouls as car travels through dips. Test car also hampered by sand tyres. Discovery had a clearance problem, too (above), dragging front spoiler as it was tugged out of mire (above). Frontera has good clearance, short overhangs, best tyres

£21,989 Shogun, now into its second generation, and aiming to be a poor(er) man's Range Rover. It shares the Frontera's front suspension layout of double wishbones and torsion bars, but at the rear its live axle is sprung by coils, and located by trailing arms and a Panhard rod. Its dampers are electronically adjustable, using a three position switch on the central armrest. The really smart parts are in its 'Super Select' 4wd system. Like the Range Rover, it features a centre differential (it's a viscous system), so can channel drive to all four wheels on the road as well as off. But unlike the British all-star, the Mitsubishi can also run in 2wd by electrically removing drive to the front wheels, a feat it can accomplish on the move. There's also a lock for the rear differential, and some colourful, illuminated graphics on the facia to give you a clue as to which of the many combinations you're in.

On paper, the Discovery's suspension layout - a live beam axle, radius arms, Panhard rod and coil springs at the front, and another live axle, trailing arms, A-frame, and coil springs at the back - looks complicated yet old-fashioned, but, as 21 years of the Range Rover prove, it all works rather well. Like the RR, the Discovery has permanent four-wheel drive, but it differs in having a manually lockable central diff instead of a viscous coupling.

As the A1 drags on towards Edinburgh, what doesn't work so well is the Discovery's power-assisted worm and roller steering. Its fuzziness around the straight-ahead leads to a disconcerting lack of stability on the motorway, and is hardly helpful when you're nipping along wiggly back roads. Furthermore, hit a deep pot-hole with the front wheels, and there's a lingering aftershock through the steering wheel. Not that the recirculating ball arrangement of the Frontera is a whole lot better, especially on the off-road rubber it's wearing for our test. There's a curious wobble through the wheel at low speed. That improves when travelling faster, but accuracy is not a word you'd readily associate with its steering. Nor with the Shogun's, another recirculating ball set-up. Such characteristics seem an inevitable penalty when a steering system has to cope with the rough and tumble of off-roading.

Heavy, barn-sized vehicles powered by turbodiesel engines don't pose much of a threat to boy racers, or even their grandparents. The 2.5-litre 104bhp intercooled Shogun has a very laid-back approach to acceleration, and motorway inclines make your foot ache with the pressure required. At least it's smooth and quiet.

Until you drive the Mitsubishi, the Frontera seems sluggardly, with practically nothing on offer below 2000rpm, which is turbo time. It's not as hushed as the Shogun, especially at idle, but the extra poke is well worth the extra din.

Decibel king is the Discovery - its 2.5-litre direct-injection turbodiesel is inherently rowdy and vibratory. The trade-off for the taxi-rank clamour is mid-range shove which leaves the other two gasping.

By the time we reach the hotel, the Discovery has impressed with the style and practicality of its cabin, but not its front seats; the Frontera has proved the most comfortable for front seat passengers, but not for those in its

89

rear; and the Shogun's cabin has revealed that the spirit of Japanese glitz is still alive and kicking.

A dark blanket of cloud is snuffing out sun rays as we meet up with Helena Bell, proprietor of the Perthshire Off-Road Centre, and her husband Sandy. They arrive in a long-wheelbase Land Rover turbodiesel, the workhorse equivalent of the boulevardiers in which we've shown up. As a farmer's daughter, Helena has been driving Land Rovers in the wilderness for far longer than she's held a driving licence, so is well placed to spot features of our dirt treaders that we might not notice.

Of the three, she's most intrigued by the Frontera. Even in long-wheelbase (109in) guise, its entry and exit angles (which determine the type of dips and slopes the car can tackle) are fine, and there are no spoilers or other low-slung clutter to impede ground clearance. It's significant that at the end of the day, the Frontera was the only one of the three not to incur damage. The Discovery's plastic front spoiler popped from its rivets when it was towed out backwards from a peaty mire by the Land Rover. Its (sturdy) tow hook is prone to digging in, too.

A long rear overhang means the Shogun thumps its rump over quite modest lumps, heavily enough during our test route to bash the moulded bumper unit out of kelter with the rest of the body. One of its optional side protection bars also suffered - kinked in the middle. And one of its rigid plastic mud flaps was left hanging drunkenly (Helena dislikes even flexible flaps; you run over them reversing out of trouble).

The further we drive the Frontera, the more surprised we are by its capabilities. We'd expected its mechanical simplicity to relegate it to being tail-end Charlie. Instead, it's chomping through obstacles even the Discovery needs a couple of bites at. Some of this is down to tyre choice, but that doesn't alter the fact that the Frontera performs its off-road duties with aplomb. What feels like a dearth of low-rev torque on the road manifests itself as a controllable trickle of muscular pull off it. The turbo's extra wallop is required only when making a frenzied assault on something steep, slippery and upwards. Downhill, there's ample engine braking and grip, even when the top surface is slimy, as it becomes when the black clouds fulfil their promise. The secret of successful off-roading is to go very slowly everywhere, and keep off the brakes. So smooth, controllable torque delivery is critical.

This is where the Discovery excels. The performance superiority it demonstrates on the road is carried over onto the dramatic Scottish hillside. The 195lb ft of torque it produces at 1800rpm overshadows the 158lb ft (at 2200rpm) of the Frontera, and the 177lb ft (at 2000rpm) of the Shogun. Ascents which require a gung-ho approach in the other two can be

Shogun has blown diesel (above) like the others, but it's not beefy enough, being slow on highway and inadequate for steep off-roading, too. It's refined, though. Discovery's direct-injection unit is noisiest but offers plenty of puff. Shogun also offers petrol V6, Discovery a V8, both of which are thirsty

cars to be supplied with mud tyres, but interpretations by the manufacturers differed greatly.

Least satisfactory were the bulbous Bridgestone Desert Duelers fitted to the Shogun. Maybe they keep you afloat on a sea of sand, but they're useless at cutting through waves of mud to get through to the more solid stuff underneath. The tread pattern quickly cakes up, turning the tyre into a slithering slick. Furthermore, they have no shoulder, so can't carve their way out of deep ruts, and slide sideways.

Land Rover supplies the Discovery with multi-purpose Michelin XM+S tyres, which look the part but don't quite measure up. Their shoulders are clear cut, the tread pattern rids itself of undesirable sludge build-up, and yet the Discovery doesn't feel as confident in places as the Frontera, whereas it should trounce it. Helena confirms, through experience, that on dedicated off-road tyres, the Discovery would give a much better showing of itself than it has on this day. The excellent BF Goodrich Mud Terrain boots Vauxhall will fit to the Frontera on demand, give it a more than fighting chance in the mud against more sophisticated opposition. It's a great combination.

But even with its tyre disadvantage, the Discovery wins out. On an equal footing, it would be over the hills and far away after a day's driving. Its engine is perfect for trekking, its chassis is bettered off-road only by the Mercedes G-Wagen's, it's reasonably well mannered on road, and it has a practical and distinctive cabin.

Don't forget how much cheaper the Frontera is, though. It, too, is very able when the track in front is made by sheep, and it's no less enjoyable to punt through the muck.

Proper tyres would improve the case for the Shogun, but I doubt that would be enough to see it overtake the Frontera. It's better off piste than most of the other contenders in the class, but simply doesn't have enough puff to compete here. Yet its refinement means most Shogun owners, who probably have no desire to take off-roading seriously, will be perfectly satisfied. If only they knew what they're missing.

Frontera's engine (right) pulls better than Shogun's, is quieter than Disco's. Shogun interior (top) well equipped, if rather flash; Frontera's (top right) comfy in front, more cramped behind; Discovery's cabin (centre) is roomy, nicely designed and practical, but front seats not comfortable enough on long hauls

attempted in the Disco with fewer revs and greater decorum.

On the flat, the Shogun wades in with all the push and shove of the others, but in a power-sapping climb, it gets wheezy and weedy. On one of Helena Bell's test hills, it stops a good 20ft short of where the second-best Frontera gets to, though it rides better - off road and on - than either of its adversaries here. But neither Shogun nor Frontera can match the amazing axle articulation of the Discovery, which helps keep four on the floor (wheels, that is), even when the ground is all ups and downs.

The role of proper tyres is often underestimated, but our day in the rough fully illustrates their importance. We asked for all three

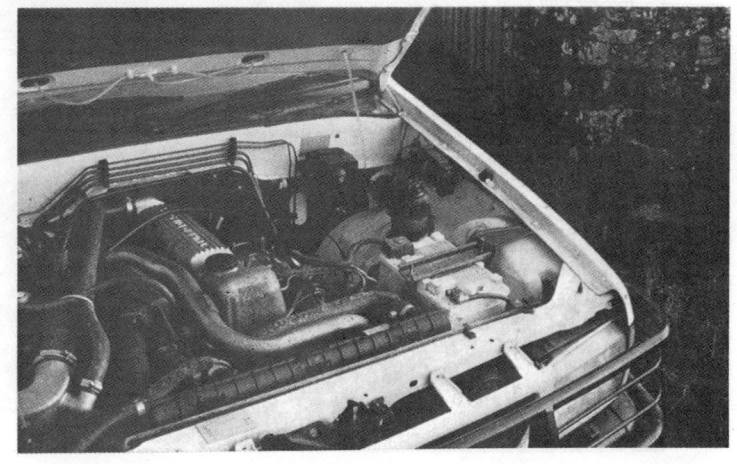

NEW ARRIVAL IN THE FAMILY

DISCOVERY V8i AUTOMATIC

Report by John Cornwall
Pics by Moray McNab

AT LAST! After waiting eight weeks from initial order, my new Discovery V8i Auto has been delivered. This to me, is an incredible amount of time to have to wait, especially in a very competitive market place sector – let's hope Land Rover will do better otherwise potential customers will start to look elsewhere. We have since learnt that the problem was due to "lack of handling kits" being available at the time my vehicle was being assembled.

You my recall in last November's LRO, my wife wrote an article enthusing over the the Auto Discovery we had on loan, well, we were due to change our Range Rover Vogue and it didn't take much persuasion from Anne to go for a new Discovery. In just over a month we have clocked up just over 2,000 miles and the vehicle has performed perfectly. In between we've had a 1,000 mile service and a phone installed – plus an elec. aerial. More on that later.

The Discovery makes quite a change from the more sober and sedate Range Rover – not quite 'executive material' (yet) but definitiely not utility Land Rover either. Its certainly taller, and with its "smoother" shape, cuts better through the air at speed – not so much a housebrick as a rounded boulder! The seats are not as plush though and I really miss the armrests (one of our regular advertisers is looking into this – hopefully he will producing some to fit as add ons, see page 44). Seating is firm, comfy and practical, but yet… no doubt Land Rover will produce an S.E. version somewhere down the one, with leather seats and trim plus ABS, all for another £2-3,000 on the price. I especially like the 2 little Dickie seats in the load area, complete with seat belts. A kiddies delight where they can make as much noise and mess as they like and fidget to their hears content, whilst making rude faces at the following cars!

Little niggles so far – lack of an elec. aerial as standard issue. For a vehicle costing £24,000 its an insult. The digital clock is placed left of centre in the dash – you have to move the head and take you eyes off the road to read it. Somebody has told me that the Discovery Interior has won a *design* award – do you think the judges confused it with *style*?

Driving the vehicle is a joy and one soon forgets the minor irritations. Superb all round vision and driving position strike one immediately. We opted for the "freestyle choice" – alloys with Michelin 235/70 R16 tyres plus anti-roll bars front and rear. They really do take out the sway and roll through fast corners. Steering is light and precise, it does go where you point it! Restricted parking areas are a doddle with the PAS taking the strain and not you, you also soon get used to the extra body length.

There is always two schools of though about autos – either you like them or you don't! Anne and I are with the former, having been auto owners for over 20 years. The ZF type 4HP22 auto 4-speed box is as per Range Rover less the chain driven transfer and VCU. It works very well and is the the quietest 4x4 auto we have ever owned, even the "traditional" transfer box whine is missing – Gulp, have we got the world's quietest Disco?

On the move, for quick mid-range speed overtaking, the usual kick-down routine is used, giving effective and smooth acceleration as desired. You can of course hold the lower gears manually as you accelerate, clicking into 'D' when your chosen speed has been reached. All very well and many drivers do it, but then why bother at all with an auto box? It seems rather self defeating.

With all this power on tap from the 3.5 V8, you obviously need effective braking. With the large diameter servo assisted discs front and rear, this is never in doubt, they are superb.

The 3500cc V8i CAT. engine produces some 152 bhp @ 4750 rpm, surely adequate for most, giving a top speed of almost 100 mph. However, a "leaden" right foot will seriously damage your wallet. We did a normal leisurely run (some motorways) from Norwich to Derby last week and averaged just on 19mpg. On that basis, the 18 gallon tank, full of unleaded, should give a 350:ISH mile range if used reasonably considering the vehicle weighs in at 2 tons (That's heavier than the Range Rover).

We haven't done any serious off roading with it yet, no doubt it will perform as well as the Vogue. With the Discovery outselling any other 4x4 in the UK, it will bring more and more newcomers into the 4WD scene. Many no doubt will never venture off-road or tow with their vehicles, preferring instead to use it as a smart, safe and reliable people carrier. Land Rover tell us that Discovery production is soon to be raised to 3000 units per month. If they can sell them and deliver in a reasonable time, they've really cracked it. Now where's the Chamois, this black paint work really does show up the marks…

USEFUL ADDITIONS

If like us, you own a dog that has to go everywhere with you; or perhaps you make frequent trips to the local refuse tip, or the garden centre etc, then carrying certain items can quickly mess up the load area. Belpar Rubber Co. have produced a high quality, ribbed heavy duty rubber load mat, that really fits the bill. With the back seats folded forward, it covers the whole rear area (even has holes cut out to poke the seat belt anchor point through). Designed for 90/110 and Discovery models, it really is a good bit of kit – simply take out and hose down after use. Cost is £64.63 for Discover. For further info contact Belpar Rubber on (0727) 856718

White collar dirt

The new Discovery MPi packs just two litres — but that's all it needs to pose in Piccadilly says Gavin Conway

TAKE YOUR AVERAGE bar room crowd. Guy over in the corner with the cowboy boots and bad-guy black jeans. Junior accountant, more than likely, and really quite a gentle sort of bloke. Wouldn't know a western bitless halter from a combine harvester, though. Or the fellow with the suit sharp enough to draw blood, eyebrows all serious business as he leans into his wafer-thin cell phone. Yep. Telephone sales. Be a while before he pays off that suit, too.

And so it goes in the image stakes, where judging a book by its cover is about as deep as it gets in a room full of strangers.

Or, as every car manufacturer capable of approximating a Mercedes grille knows, on a road full of strangers.

That brings us rather neatly to the newly crowned king of the urban pose.

The 4x4 market is growing explosively; it expanded by 35 per cent last year and upwards of 55,000 of the cars will find homes in the UK this year.

Now, as any outdoorsman worth his woodpile will tell you, that adds up to a whole lot more off-roaders than we've got off-road.

Which is just fine, because like the accountant/cowboy, most off-roaders won't even get dusty, let alone down and dirty.

Enter the off-road sales king, Land Rover, and its latest, greatest, evolutionary two-litre petrol-powered Discovery MPi. Evolutionary not because it is a better off-roader — it packs the least torque (137lb ft) of any

Discovery. And changing lanes in big taxi country is great fun, too. Just indicate and slowly move over; even the most obnoxious cabby realises that a Discovery changing lanes is as irresistible as continental drift. Head and legroom? Don't make me laugh. Men with hats will have headroom to bounce with.

Parking the beast is stupidly easy, too. It is actually shorter than a Granada. And with its prodigiously boosted power steering, the Discovery's fat Michelin 235/70R15 tyres can be swivelled a good deal easier than those of even a Fiesta.

Great in traffic, easy to park, bags of shopping room. Could this be the car to rival the Cinquecento and Twingo?

Well, no. First off, it is still rather a big tool. And compared with newcomers such as the Jeep Cherokee and Ford Maverick, the Discovery is a less compromised off-roader. Which means it isn't quite so good on the road.

But the crux of this Discovery is the two-litre engine. Frankly, it needs more heart.

The 2.5-litre turbo diesel Discovery growls and cackles and generally makes enough macho noise to round out the hard-man-comes-to-the-city image nicely. You can even rock it at the lights by blipping the throttle, connected as it is to 195lb ft of torque. The V8 Discovery needs no more credibility than its badge. But a two-litre Disco? Isn't that the same engine as the one in my 820SLi? How humiliating.

The Discovery MPi does regain a little macho ground, though, with the noise it makes. Its engine is coarse and thrashy and when pushed past 4000rpm begins to sound a lot more like its ballsy and boomy turbo diesel brother.

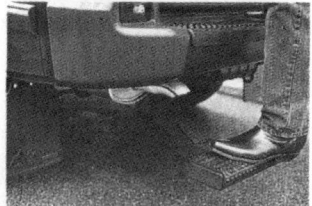

On top of the world in town

Steppin' out: rear access

It is slow, too. Progress is not quite as funereal as in the TDi, but 0-60mph in a claimed 15.3secs and a top whack of 98mph ain't great. Wheedling maximum performance out of the engine is positively painful, so a subdued yet majestic driving style works best. Drive this Disco even a little perkily and you'll be thrashed at the fuel pumps. I managed an average fuel consumption of just 16.5mpg in combined motorway and city motoring; Land Rover claims an urban figure of 18.4mpg, 29.2 at 56mph and 21.2 at 75mph.

But — and this is an important but for those who want the Discovery cachet but don't have the cash — the three-door £16,995 Discovery MPi is the least expensive model next to the £17,495 Discovery TDi.

I like it. I like the phenomenal space, the driving position and the front-row pecking order afforded by nearly two tonnes of attitude. I'm not crazy about the lethargic powertrain, though, but I like the rough guy looks.

Oh, and the black jeans and cowboy boots pictured above belong to me. Enough said. ■

Engine from Rover 800

Plenty of usable space

Discovery — but because of its admission that urban cowboys don't really need huge masses of torque to haul them up the two per cent gradient into Sainsbury's.

Land Rover says it "has identified a group of customers who may not need the heavy duty towing and ultimate off-road performance of the existing TDi and V8 engines".

Brilliant! An off-roader aimed specifically at people who don't need an off-roader!

The Discovery MPi sports the same 134bhp Rover T16 engine that currently serves the Rover 800 range. Driving the point of its new concept home, Land Rover says that it hopes to entice company car drivers out of their Senators and Granadas and into a Discovery.

But should a hapless driver raised on motorway mileage make the wrong turn and find himself in *Twin Peaks* territory, the Discovery, apart from its rep-motor transplant, is still every inch a serious mud-mobile. From its huge 206mm (9ins) ground clearance to its massive amount of wheel travel, the Discovery remains capable of truly big adventures.

In fact, many of the Discovery's serious off-road credentials are exactly what make it such neat fun as a city car. Its Olympian driving position lets you know about trouble brewing in the next county, while most other drivers' view will be completely blocked by anything bigger than, say, a

△ *This automatic Tdi Discovery has 235x70 tyres on five spoke alloys and a handling kit*

Report and photos: Nick Dimbleby

Auto Tdi

IT'S GOOD to see that at last Land Rover seem to be taking notice of what their customers want. After taking thirteen years to launch a four-door Range Rover, and just over thirty to bring out a coil sprung Land Rover, the company hasn't exactly had a good track record in the past. However, with the launch of the Discovery Tdi automatic and the enlargement of the 3.5 V8i to 3.9 litres earlier this month, Land Rover are beating the Japanese at their own game - they're producing the vehicles for which customers are asking.

By launching the Discovery Tdi automatic, Land Rover is hoping to take customers away from the two-wheel drive 'executive' diesel car sector, by convincing them that Discovery is a 'car' that can combine business with pleasure.

Apparently, market research has shown that four groups are most likely to buy the Tdi automatic: current Discovery owners who prefer life without a clutch; owners of other 4x4s who rejected the Discovery for its heavy clutch; owners of two wheel drive diesel automatics who are attracted to Discovery's versatility; and prospective customers who had previously rejected the Discovery Tdi because it lacked an automatic option. If their research is right, Land Rover estimate that up to 25% of future Discovery Tdi sales will be without a clutch.

Behind the wheel

Slipping behind the wheel of the Discovery Tdi auto, I noticed that thicker carpets are now standard on all models; and after starting the engine, I was surprised to find how much quieter and more refined the 1994 Discoverys are. This is due to improved soundproofing under the engine and on the transmission tunnel, as well as a few minor details like revised rear seatbelt mounts and the thicker carpets. If you're used to the shakes and rattles of the old diesel Discovery, take a test drive in the new model: I think you'll be impressed.

All of the Discovery Tdi press fleet were equipped with the new 'Freestyle Choice' option pack, that consists of specially developed 235 x 70 tyres on 16 inch five spoke alloys, along with front and rear anti-roll bars to stop the lean. The new wheels and tyres look good, and there can be no doubt that the handling kit improves the body roll around corners. We hacks also enjoyed the six-disc CD player with extra sub-woofer and the factory-fitted front and rear air conditioning. Both these items are optional, but well worth going for.

Clicking the familiar ZF four-speed box into drive, I headed out onto the open road. The research and design team have reprogrammed the gearbox so that the box clicks into top at 47 mph, as opposed to 55 mph on the automatic V8 Range Rover. Like the Range Rover, there is a kickdown facility, although comparisons to the V8 are cruel - the Tdi just doesn't have the top end

△ *Tdi engine in Discovery has optional air conditioning*

△ *It's automatic and it's diesel*

◁ *3.9 V8 in the Discovery*

torque to briskly move the needle up the speedometer.

However, the Tdi auto is more than sufficient for 'normal' motoring; you just have to be aware that you can't floor it in the same way as you can with a V8. On the other hand, it's unlikely that potential buyers of a Tdi Discovery are looking for a car with bristling performance; they're much happier with the vehicle's superb fuel consumption: some 28.2 miles per gallon in the simulated urban cycle.

Apart from a good blast around a recently harvested corn field, we were unable to test the Tdi automatic off-road (they mentioned something about scratching the paint…). However, we did engage low box to make sure that everything worked, and this revealed that engine braking is much better than the V8 auto.

Currently, diff. lock has to be engaged manually by the driver, and Land Rover have said that there are no plans to install the VCU transfer box as used in the Range Rover. However, it's probably only a matter of time before this well-proven unit gets transplanted into the Discovery.

All in all, the Discovery Tdi automatic should sell very well. It fills a gap in both the model range and the market, and is reasonably priced at £1150 more than the manual equivalent. Place your bets now for the length of time it takes to launch a Range Rover Tdi automatic.

More V8 power

The unveiling of the 1994 model year Discovery also sounded the death knell of the seven year old 3.5 litre V8 Efi powerplant. From now on, only 3.9 litre V8s will be produced at Solihull for Range Rovers, Discoverys and US-spec. Defenders.

The extra 400 cc gives a 20% increase in power and torque, and the 3.9 litre manual I drove certainly seemed to have plenty of get up and go. The larger engine is smoother and quieter than the original, making the V8 3.9 Discovery a very desirable alternative to a Range Rover. However, Range Rover sales are unlikely to decline, as the Discovery image appeals to a very different market.

The Discovery V8 is mainly sold abroad, and makes up approximately 20 - 30% of total sales. Main markets include Australia, Japan, Switzerland and the Middle East, where petrol engined cars are particularly popular. In the home market, the 3.9 litre Discovery should help see off competition from the 4.0 litre Jeep Cherokee and the 3.1 litre Isuzu Trooper.

The continuing development of Discovery (as well as the announcement of two new colours) should ensure that the vehicle continues to be the market leader in the leisure 4x4 sector for some time. I just hope that Land Rover's future marketing and development strategy continues in such a similarly effective vein.

new CARS

Land Rover Discovery Mpi

Street cred: this is new 2.0 petrol Discovery's more natural habitat – short of pulling power for serious off-roading

LESS IS MORE FOR NEW DISCOVERY

Stick to the tarmac and the cheaper, 2.0-litre Discovery is a sensible choice

Firing up the latest Discovery can come as a shock, especially if you've driven a diesel or V8 petrol version in the past.

For instead of the wonderful V8 burble or diesel clatter, the 2.0-litre, petrol-powered Mpi offers the quiet hum of a family saloon. But that's not the only shock. The 3dr Mpi Discovery is now the cheapest version of the Land Rover you can buy. At £16,995, it undercuts the diesel equivalent by £500 and the 3dr petrol V8 version by almost £1200.

Using the 134bhp 2.0-litre Rover engine otherwise found in the 400- and 800-Series, the Mpi will be available in both three-door and five-door bodies, the five-door in both basic and 'S' trim. And it's the 5dr S, tipping the scales at 1925kg (that's equivalent to two five-door Fiestas) and costing £19,900, that we drove for this test.

DRIVING THE Mpi
KEEP OFF THE GRASS!

On the face of it, mating a 2.0-litre engine to a Discovery seems pretty daft. Sure, the Mpi's 0-60mph time of 15.3sec may actually be better than the 200 Tdi 5dr's 16.2sec, and the 2.0-litre's 134bhp outguns the diesel's 111bhp.

However, what's needed more than good sprinting ability is low-down pulling power, especially for off-roading. It's here that the Rover unit falls short – compare the 137lb ft of torque generated by the small petrol engine with the hefty 195lb ft churned out by the diesel.

But do the figures betray themselves on the road? It all rather depends upon which roads you're on. Keeping up with the flow of traffic around town involves plenty of gearchanges, and the engine needs to be worked fairly hard to keep pace. That explains the official 18.4mpg figure for urban driving, not much better than the V8's 15.2mpg. Acceleration is acceptable, though.

On the motorway, the Mpi will also cruise happily, although we found that a change down from fifth was needed on inclines to maintain speed – acceleration in top gear from 50 to 60mph and beyond is like watching a kettle boil.

We didn't get the chance to drag the Mpi off-road but, until we do so in a full test, we would venture to suggest it should never face anything more serious than the odd run across a field. There simply isn't the pull in high gears or the tractability to get it out of really serious trouble.

One of the drawbacks of this big off-roader is – still, despite claimed improvements – the quality of the gearchange, especially from first to second. It's just plain nasty, and needs patience if it isn't to snatch in your hand. You *can* drive around it after a while, but why should you need to?

SPACE AND ACCOMMODATION
HOW BIG A FAMILY DO YOU NEED?

The five-door Mpi makes a mockery of the word 'cavernous'. Headroom and legroom both front and back are generous to a fault, while storage space for oddments is marvellous. Flipping down the two side seats at the rear turns the Discovery into a seven-seater; folding the rear seats gives enough room to load just about anything you can throw at it. What more could you ask?

WHY BUY AN Mpi?
TAX ADVANTAGES; SEVEN SEATS

In fairness to Land Rover, the Mpi is aimed at a very specific buyer. He or she will probably be moving up from a 2.0-litre estate or family saloon, and many will be viewing it as a company car (the Mpi offers significant tax advantages over the V8).

All potential customers *will* be looking for the traits that have made the Discovery successful – sheer carrying capacity and a lofty view of the road. They'll more than likely not care about the comparative shortage of power and will probably never venture this Land Rover anywhere near a rocky hillside or sand dune. If all that sounds like you, take a look at the Discovery Mpi.

FIRST GLANCE
VERDICT ●●●●○

No, Land Rover hasn't performed a miracle with the Mpi. The 2.0-litre engine's shortage of clout relegates it to on-roading and very mild off-roading. But it still has all the Discovery plus-points – tons of space and a 'King of the Road' feel. And for company drivers, it offers some serious tax advantages over its bigger brothers.

LAND ROVER DISCOVERY Mpi

ENGINE 1994cc 4cyl, 134bhp
PACE 98mph, 0-60mph 15.3sec
TOURING MPG 21.7
PRICE £16,995 (3dr); £18,450 (5dr); £19,900 (5dr S)

ON SALE Now

BEST RIVAL

Jeep Cherokee 2.5 Sport, £15,995: Cheaper, quicker, better specified. But driving position is not as commanding and there's nothing like the same storage space.

S-badged version of Mpi comes with electric windows and alloys...

...plus all the Discovery's virtues of bags of space, front and rear

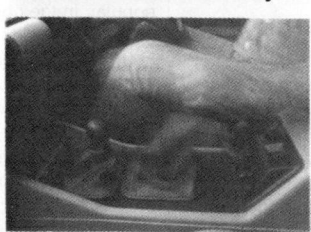

Notchy, imprecise gearshift is still a Land Rover Discovery drawback...

...but Discovery pluses of plentiful storage space, neat design retained

ROAD TEST

By James Taylor

△ Discovery Mpi

Urban cowboy

James Taylor tries out the new small-engined Discovery. It's a pleasant enough vehicle, but ...

LAND ROVER know only too well that they could sell even more Discoverys than they already do if only they had an engine which wasn't either noisy (like the Tdi) or thirsty (like the V8).

Their market research tells them that there's a whole group of potential customers out there, currently driving conventional 2-litre saloons and estates, who would give their eye teeth to have the image, space and perceived strength of a Discovery.

They aren't kidding, either: I know quite a lot of people like that myself.

So, just as in the old days, Solihull's engineers have borrowed an engine from a Rover saloon car and made it fit their own vehicle. The engine is the 2-litre, twin-overhead camshaft four-cylinder which was introduced last year in the second-generation 800-series saloons, and it goes under the name of T16.

In the cars, it's mounted transversely with an end-on gearbox, and so to fit Discovery it had to be adapted for a north-south installation. Fuelling is by a multi-point injection system, and it's that which gives its initials of Mpi to distinguish the leaflet version of the Discovery from its brothers.

The T16 isn't quite as new as you might think, however. It's actually a redeveloped M16 engine, and the M16 appeared as long ago as 1986 in the first-generation Rover 800 series cars. Bore, stroke, belt-driven twin overhead camshaft configuration and iron block/alloy head construction are all carried over.

What's new about the T16, however, is it's improved induction system, which gives much better low speed torque, and the eight counterweights on its crankshaft, which give smoother running than the M16's four counterweights. The T16 looks a lot neater, too.

Whereas the M16 is festooned with external breather pipes, the T16 has breather channels cast into its block and heads. As a result, it's also simpler to manufacture than the M16 and less prone to leaks.

The T16 has gathered itself a good reputation in the 800-series and, more recently, in the Rover 220GTi coupe. The M16, by comparison, wasn't so well regarded – mainly because of teething troubles with early production examples. So the engine comes to the Discovery with a good track record. It also comes with a power output of 134 bhp which, on paper, puts it nearly between the 153 bhp V8i and the 111 bhp Tdi.

However, things aren't quite that simple. For a start those 134bhp are developed at a screaming 6,000rpm, which means that most owners probably won't use the Mpi's 98mph maximum speed for prolonged periods. In a heavy vehicle like the Discovery, it's torque which counts for much more than power. The Mpi's 137 lb/ft at 2,500rpm begins to look a little sick, even on paper, alongside the Tdi's 195 lb/ft at 1,800rpm and the V8i's 192 lb/ft at 3,000rpm.

So it proves on the road. The Mpi gets left for dead at the lights by almost anything dri-

T16 engine in the Discovery Mpi

ven with spirit. Acceleration definitely isn't its strong point, and you need to be prepared to use the gears a lot, particularly in hilly country.

Once on the move, it proves to be reasonably refined. However, our test vehicle suffered from a collection of underbonnet zizzes and zings which suggested that something hadn't been sorted properly.

The Mpi bowls along at a cracking pace, but you quickly recognise that there isn't enough urge to start overtaking on fast A-roads unless you can see a long way ahead. The vehicle's inability to make progress in these situations can be mite frustrating. Fuel economy in the low 20s isn't all that much of a compensation, either.

That lack of torque also means that the Mpi is not the right choice of Discovery if you intend to do any serious towing. Land Rover quote the maximum towing weight as 2,750kg, as compared to 4,000kg for the bigger-engined models. Those figures tell their own story.

As far as off-roading, the Mpi just doesn't inspire enough confidence to encourage drivers to take it anywhere more demanding than a muddy car park. It's worth betting that there will be a lot of Discovery Mpi models on the market in five years' time with transfer boxes which have never been used in low ratio.

Mind you, that says as much about the sort of person who will buy the Mpi as it does about the vehicle's own abilities. The press demonstrator which Land Rover provided for our test was a five-door Mpi S, complete with the extra-cost Freestyle Choice package of a star-pattern alloy wheels, fatter 235/70 tyres, and front and rear anti-roll bars. It's undeniably stylish, and it's aimed at those who want a stylish vehicle. Put it up against a Volvo estate or a Renault Espace, and there's just no comparison on looks. The Discovery wins hands down.

For those who might have been in the market for a Volvo or an Espace, the Discovery's four-wheel-drive will be seen as an aid to on-road traction. Its robust construction will be seen as an accident-safety factor, but its Land Rover ancestry will be seen as little more than a designer label. That's a great pity but, if it sells more Land Rover products and keeps the company healthy, perhaps we shouldn't complain.

To summarise, the Discovery Mpi is a very pleasant vehicle, with all the style, robustness and space of its larger-engined brothers. Any owner of a 2-litre estate car who tries one out is likely to be very impressed. The people who won't be so impressed are those who know just how good the Tdi and V8i models are. To them, the 2-litre Discovery will come as a disappointment.

There is, by the way, a turbocharged version of the T16 2-litre in production. It's fitted to the latest 800-series Rover Vitesse and to the 220 Turbo Coupe. The Vitesse offers a useful 180bhp, although its peak torque is delivered at a sky-high 4,000rpm.

In the coupe, on the other hand, the engine has been tuned for 197bhp at 6,100rpm and 175lb/ft of torque at just 2,100rpm.

Now wouldn't you love to see one of those in a Discovery?

Range Rover Gold Portfolio 1979-1992

This updated edition of our earlier Gold Portfolio on the Range Rover now contains some 50 articles. Included are road and comparison tests, technical data, a 'used survey', new model introductions, feed-back on rallying, long term and off-road reports, information on touring plus full specifications and advice on 'buying second-hand'. All models are covered including the Schuler, Vogue, Automatic, Turbo Diesel and the Janspeed.

180 Large Pages - 380 illustrations